Molecular Pharmaceutics

Nano Technology and Targeted
Drug Delivery Systems

Molecular Pharmaceutics
Nano Technology and Targeted Drug Delivery Systems

Dr. Suryakanta Swain
Dr. Sarwar Beg
Dr. Rabinarayan Parhi

WOODHEAD PUBLISHING INDIA PVT LTD

New Delhi

Published by Woodhead Publishing India Pvt. Ltd.
Woodhead Publishing India Pvt. Ltd.,
303, Vardaan House, 7/28, Ansari Road,
Daryaganj, New Delhi - 110002, India
www.woodheadpublishingindia.com

First published 2019, Woodhead Publishing India Pvt. Ltd.
© Woodhead Publishing India Pvt. Ltd., 2019

Woodhead Publishing India Pvt. Ltd. ISBN: 978-93-88320-04-7
Woodhead Publishing India Pvt. Ltd. e-ISBN: 978-93-88320-03-0

Typeset by Bhumi Graphics, New Delhi
Printed and bound by Replika Press Pvt. Ltd.

Contents

Preface   *ix*

Foreword   *xi*

Acknowledgements   *xiii*

List of Contributors   *xv*

**1. Targeted Drug Delivery Systems: Concepts, Events and   1
     Biological Process Involved in Drug Targeting, Tumour
     Targeting and Brain-Specific Delivery**

    1.1   Introduction   1

    1.2   Concepts of Targeted Drug Delivery Systems   2

    1.3   Objectives of Targeted Drug Delivery Systems   3

    1.4   Ideal Characteristics of Targeted Drug Delivery   3

    1.5   Advantages   3

    1.6   Disadvantages   4

    1.7   Classifications of Drug Targeting   4

    1.8   Components of Targeted Drug Delivery   4

    1.9   Strategies of Drug Targeting   7

    1.10 Biological Processes and Events Involved in Drug Targeting   9

    1.11 Tumour Targeting and Brain-Specific Drug Delivery   14

    1.12 Conclusions   18

    1.13 References   18

    List of Abbreviations   20

**2. Targeted Methods: Nanoparticles and Liposomes: 21
     Types, Preparation and Evaluation**

    2.1   Introduction   21

    2.2   Classification of Nanoparticles   22

    2.3   Production of Drug Nanoparticles   23

    2.4   Characterization of Drug Nanoparticles   25

    2.5   Therapeutic Applications of Nanoparticles   28

    2.6   Classification of Liposome   30

    2.7   Materials Used in Liposomes   31

2.8   Liposome Preparation Methods   32
2.9   Characterization of Liposomes   34
2.10   Therapeutic Applications of Liposome   36
2.11   Conclusions   37
2.12   References   37

**3. Microcapsules/Microspheres: Types, Preparation and   41
      Evaluation; Monoclonal Antibodies; Preparation and
      Application, Niosomes, Aquasomes, Phytosomes,
      Electrosomes; Preparation and Application**

3.1   Introduction   41
3.2   Micro Capsules/Micro Spheres   43
3.3   Monoclonal Antibodies   60
3.4   Niosomes   72
3.5   Aquasomes   83
3.6   Phytosomes or Herbosomes   89
3.7   Electrosomes   94
3.8   Conclusions   94
3.9   References   95
       List of Abbreviations   109

**4. Pulmonary Drug Delivery Systems: Aerosols, Propellents,   111
      Containers Types, Preparation and Evaluation, Intranasal
      Route Delivery Systems; Types, Preparation And Evaluation**

4.1   Introduction   111
4.2   Anatomy of Lung   111
4.3   Aerosol   112
4.4   Propellants   113
4.5   Containers   116
4.6   Valve and Actuator   118
4.7   Product Concentrate   121
4.8   Homogenous Aerosol Systems   121
4.9   Heterogeneous Aerosol System   122
4.10   Manufacturing of Pharmaceutical Aerosols   124
4.11   Testing of Pharmaceutical Aerosols   126
4.12   Biological Testing   130

4.13   Intranasal Route: Nasal Drug Delivery System   130

4.14   Conclusions   133

4.15   References   133

List of Abbreviations   133

5.   **Nucleic Acid Based Therapeutic Delivery System: Gene   135
     Therapy, Introduction (Ex vivo and it is In vivo Gene Therapy),
     Potential Target Diseases for Gene Therapy (Inherited Disorder
     and Cancer), Gene Expression Systems (Viral and Nonviral Gene
     Transfer), Liposomal Gene Delivery Systems, Biodistribution and
     Pharmacokinetics, Knowledge of Therapeutic Antisense Molecules
     and Aptamers as Drugs of Future**

5.1   Introduction   135

5.2   Types of Gene Therapy   137

5.3   Different Vectors Used For Gene Delivery   140

5.4   Liposomal Gene Delivery Systems   144

5.5   Biodistribution and Pharmacokinetics, Knowledge of   144
      Therapeutic, Antisense Molecules and Aptamers as
      Drugs of Future

5.6   Preclinical and Clinical Trials   150

5.7   Conclusions   151

5.8   References   151

List of Abbreviations   153

Index   155

Preface

It is our prime intention to cover the chapters of this series as comprehensively as possible. Thus, we are very pleased to introduce this textbook "MOLECULAR PHARMACEUTICS (NANOTECHNOLOGY AND TARGETED DDS) (NTTDS)". The present book is unique in several aspects. It provides a map of the body from the viewpoint of drug targeting, drug delivery vehicles and different drug delivery systems. We are extremely grateful to all the contributors in this book, who have given up encouragement to write about all the selected chapters of molecular pharmaceutics and targeted drug delivery systems to create imagination in the mind of B. Pharm, M.Pharm students, Pharm D and Ph.D scholars. Preparing this book took longer than anticipated, and it contains more pages than expected. This present book should prove to be useful to pharma students studying in the Pharmaceutics, Industrial Pharmacy, and Pharmaceutical technology specialization, and could help to develop their career in the field of pharma industries as Formulation R&D scientists, technicians, etc. I would like to think that this book might fit any of the above description, depending on the reader's need. I heartily thankful to my both the co-editors Dr. Sarwar Beg and Dr. Rabinarayan Parhi for their constant and keen involvement in compilation, edition as well as creation of flow charts and figures of few selected chapters during writing of this book. Finally, I would like to thank my loving wife **Linarani Swain** for her love, understanding and constant support during the time of preparation of this textbook. I would like to thank my loving son **Priyans Swain** for giving me some time to make this book.

Dr. Suryakanta Swain

Foreword

There are very limited textbooks covering the most of the advanced aspects of **Molecular Pharmaceutics (Nano Technology and Targeted Drug Delivery Systems)**. A need for such a book written in a simple, direct, lucid and relevant language was appropriately realized. The excellent efforts of Dr. Suryakanta Swain and Co-Editors Dr. Sarwar Beg and Dr. Rabinarayan Parhi have resulted in a faithful publication of this textbook covering major area related to concepts, events and biological process involved in drug targeting, tumour targeting and brain-specific delivery in academic and industrial pharmaceutical research labs. The subject matter is written with adequate practical tips and theoretical background and presented in a simple manner for better understanding among the M. Pharm., B. Pharm., Pharm D students and Ph.D research scholars. I hope that you as a reader, whether you are a student, teacher in india or abroad, researcher, academicians would be pleased for this text book, which should certainly add to your current knowledge, understanding and insights of theoretical and practical concepts of this selected topics covered.

It is a matter of immense pleasure for me to write a foreword for this textbook. I wish the editor and co-editors and all the contributing authors and coauthors of all the chapters a great success in this combined venture and hope that their contribution to pharmaceutics or pharmaceutical sciences literature will contribute endlessly.

The chief-editor of this textbook, Dr. Suryakanta Swain, is a well-established scientist with extensive academic teaching and research knowledge in the field of pharmaceutics and pharmaceutical sciences. Dr. Suryakanta Swain has been doing research in pharmaceutical targeting drug delivery systems for last 11 years, and has patented and published articles of research and review, invited editorials or opinion articles, short communications, thematic issues, international books and book chapters in reputed publishers.

Prof. Rama Rao Nadendla
M. Pharm, Ph.D., FIC.,
Dean Faculty of Pharmacy-Acharya Nagarjuna University,
Principal, Chalapathi Institute of Pharmaceutical Sciences (Autonomous)
Chalapathi Nagar, Lam, Guntur 522034, Andhra Pradesh

Acknowledgements

I acknowledge with grateful appreciation, the major contributions of co-editors Dr. Beg and Dr. Parhi in sustaining the vitality of this textbook. Their respective expertise in the fields of targeted drug delivery systems has allowed the integrated approach utilized in this textbook. I extend my gratitude with appreciation to all the contributing chapters' authors or co-authors and academic colleagues, industry friends and my well-wisher who have shared their thoughts with me. I especially thank the publisher, acquisitions editor, managing editor, copy editor and production manager of Woodhead Publishing India, New Delhi contributed so expertly to the planning, preparation and production of first edition of this textbook. Finally, I acknowledge to my sweet wife Mrs Linarani Swain and my lovable son Priyans Swain supported me lot during my entire assignment.

Dr. Suryakanta Swain
Editor

List of Contributors

Dr. Amaresh Prusty
M.Pharm, Ph.D.
Asst. Professor
Department of Pharmaceutics
College of Pharmaceutical Sciences
(Affiliated to Biju Patnaik University
of Technology)
Puri-752002, Odisha, India
E-mail: amareshprusty@gmail.com

Dr. Rabinarayan Parhi
M.Pharm, Ph.D.
Assistant Professor
Gitam Institute of Pharmacy
GITAM University (Deemed to be
University), Gandhinagar Campus,
Rushikonda
Visakhapatnam-530045, Andhra
Pradesh, India
E-mail: rabi59bls623@gmail.com

Dr. Satya Prakash Singh
Ph.D., M. Pharm
Associate Professor
Department of Pharmaceutics
Integral University, Lucknow
Pin-226026, (INDIA)
E-mail: singh.satyaprakash@
rediffmail.com

Dr. Chinam Niranjan Patra
M. Pharm, Ph.D.,F.I.C
Professor in Pharmaceutics,
Department of Pharmaceutics
Roland Institute of Pharmaceutical
Sciences
P.O: Khodasingi, Berhampur (Ganjam)
Pin-760010, Odisha (INDIA)
E-mail:drchniranjanpatro@gmail.com

Dr. Kahnu Charan Panigrahi
M. Pharm., Ph.D
Assistant Professor
Department of Pharmaceutics
Roland Institute of Pharmaceutical
Sciences
P.O: Khodasingi, Berhampur (Ganjam)
Pin-760010, Odisha (India)
E-mail:kanhu.pharma@gmail.com

Dr. Suryakanta Swain
M. Pharm, Ph.D., FIC.,PCPV
Associate Professor-cum-Head
Department of Pharmaceutics
Southern Institute of Medical Sciences
College of Pharmacy, SIMS Group
of Institutions, Mangaladas Nagar,
Vijayawada Road, Guntur522 001,
Andhra Pradesh, INDIA.
E-mail:swain_suryakant@yahoo.co.in

Dedication

- *I dedicated to my Parents, Wife and Son because they gave me strength, focus of heal and courage to share my pharmacy knowledge for successful completion of this textbook.*

- *I dedicate this textbook to my all well-wishers who always encourage me an achieving higher goal and motivating me for writing this type of textbook for Pharmacy Education and Research.*

1

Targeted Drug Delivery Systems: Concepts, Events and Biological Process Involved in Drug Targeting, Tumour Targeting and Brain-Specific Delivery

Amaresh Prusty[1*], Siva Prasad Panda[2],Suryakanta Swain[3]

[1] Department of Pharmaceutics, College of PharmaceuticalSciences, Puri-752002, Odisha, India

[2] KL College of Pharmacy, KL University (Deemed to be University), Green Fields, Vaddeswaram, Guntur-520002, Andhra Pradesh, India

[3] Southern Institute of Medical Sciences, College of Pharmacy, Department of Pharmaceutics, Guntur-522001, Andhra Pradesh, India.

1.1 Introduction

Generally, over the past 50 years large number of studies has been made in management of disease by modifying the function and property of drugs where a drug can be delivered to its targeted area at a rate and concentration that both minimize side effects and maximize therapeutics effect where drug will be maximally beneficial to the patient. Since from last few years, there is tremendous development in controlled drug delivery technology, which leads to the development of various clinical formulations improving patient compliance and convenience. These developmental techniques allow delivery of drugs at desired release kinetics extending drug release over extended period of time. Oral and transdermal drug delivery systems routinely deliver drugs for 24 h, thereby improves drug efficacy and minimize side effects. The term "targeted drug delivery" (or "drug targeting") used in drug delivery and "targeted therapy" (or "targeting therapy") are two terms with different meanings. Targeted drug delivery explains drug accumulation within a target zone but it is independent of the method and route of drug administration (Torchilin, 2000). Although the targeted therapy or the targeted medicine cites specific interaction between a drug and its receptor at the molecular level (Gerber, 2008; Mimeault et al., 2008). To examine the drug utilization, it is necessary to deliver drug to its target tissue in correct amount at a proper time to elicit the desire response. Moreover, the drug delivery must be continuous at a rate such that condition in question is cured or controlled in a minimum time with fewest side effects (Strebhardt et al., 2008).Targeted drug delivery releases medicaments in the tissues of interest by reducing the relative concentration of the medication in the remaining tissues. To qualify as 'druggable', a target must be accessible to the proposed drug molecule, and a measurable biological reaction must be provoked because of the drug interacting with the target. This reaction may be measured both *in vitro* and *in vivo*. Hence, in 'a Targeted drug delivery system, medicament is selectively targeted or delivered or shows its pharmacological action only to its site of action but not to the non-target organs or tissues or cells of our body (Mills et al., 1999). After the effective transmission by carriers, drugs are concentrated and located in the said tissues, organs, cells or cell structures, which make the drug concentration on particular area higher than that of other normal parts so that the therapeutic effects are enhanced and the toxic side effects are reduced .With the rapid progress in cell biology, material science and molecular biology, the targeted agents have gotten rapid development, which provides a platform among experts on pharmacy and pharmaceutics in various countries to study further in this field. Targeted agents in the field of western medicine have been extensively studied and applied clinically. At present, in Europe and the United States and other developed countries, related products have been available (Vyas et al., 2002). The targeting role of drugs is achieved mainly by vector system, so researching and profoundly understanding the structural characteristics and mechanism of drug carrier are the basis to achieve clinical targeting drug delivery system. In addition, more attention should be paid to the research on the carrier materials in order to develop more suitable carriers, which can reduce the toxicity of drugs, enhance their activity and improve the overall quality of all targeted agents. It is necessary to strengthen the modification of the existing drugs to make them transferred from passive targeting to active targeting, hence to improve the treatment effect (Brahmankar et al., 2009).

1.2   Concepts of Targeted Drug Delivery Systems

Paul Ehrlich, who was a microbiologist in the year 1906 first introduced the concept of targeted delivery system and proposed the idea of drug delivery. He introduced a term for targeted drug delivery and named it as magic bullet, but the main drawback was finding the proper target for a particular disease state and also finding a drug that effectively treats a disease. It also has drawback in finding a means of carrying the drug in a stable form to specific sites by avoiding the immunogenic and nonspecific interactions that efficiently clear foreign material from the body. Targeted drug delivery not only releases pharmacologically active moiety to the desired target with optimum therapeutic concentration for desired pharmacological response, it also restricts its access to normal cellular lining by minimizing therapeutic index. The drug can be targeted to intracellular sites, virus cells, bacteria cell. The targeted drug delivery minimizes drug distribution in the non-target cells but increases higher and effective concentration at the targeted site and maximizes the benefits of targeted drug delivery (Bhupinder Singh et al., 2011). Drug delivery to the body occurs by two approaches such as local and systemic ways. Local delivery of drugs release drugs to the external sites of the body while drug delivery to internal sites of the body belongs to systemic drug delivery. In case of systemic delivery drugs with suitable therapeutic concentration of drug reached the majority body parts and maintenance of this therapeutic concentration requires a large dose of the drug which may leads to various side effects. Hence, to avoid these problems associated with the conventional systemic delivery of the drugs, there is a need for the development of a targeted drug delivery system which can deliver the drug selectively to the diseased site in a specified steady concentration for the prescribed time (Ehrlich, 1960).

1.3   Objectives of Targeted Drug Delivery Systems

In order to achieve a desired pharmacological response at selected body sites without having any undesirable interaction at other sites, the drug must have a specific therapeutic action with minimum side effects (Gupta et al., 2011). Targeted drug delivery is a special form of drug delivery system where the pharmacologically active agent or medicament is selectively targeted or delivered only to its site of action, but not to the non target organs or tissues or cells.

1.4   Ideal Characteristics of Targeted Drug Delivery

- The preparation of the delivery system should be easy or reasonably simple, reproductive and cost effective.
- It should be nontoxic, biocompatible, and biodegradable.
- Restrict drug distribution to target cells or tissues and should have uniform capillary distribution.
- It must have physicochemical stable *in vivo* and *in vitro* property.
- It shows controlled drug release kinetics.
- Drug release does not affect the drug action.
- It should have the property by releasing therapeutic amount of drug in the targeted area.
- Carriers used in targeted drug delivery must be biodegradable and easily eliminated from the body (Bhargav et al., 2013).

1.5   Advantages

- Toxicity is reduced by delivering a drug to its target site, thereby reducing harmful systemic effects.
- Drug can be administered in a smaller dose to produce the desire effect.
- Avoidance of hepatic first pass metabolism.
- Dose is less compared to conventional drug delivery system (Bhargav et al., 2013).

1.6   Disadvantages

- Rapid clearance of targeted systems because as we know drug clearance is concerned with the rate at which the active drug is removed from the body. Therefore, in targeted drug delivery the particular targeted organ clears drug more rapidly.
- Immune reactions against intravenous administered carrier systems.
- In tumour cells, it shows insufficient localization, which is a major disadvantage of targeted drug delivery.
- It requires skill manufacturing drug delivery preparation and highly sophisticated technology for the formulation.
- Drug deposition at the target site may produce toxicity symptoms (Bhargav et al., 2013).

1.7   Classifications of Drug Targeting

When a new drug or chemical entity was discovered, it often posses some challenges to scientist to treat it as drug like, it may possess poor solubility problems or insufficient in vitro stability (shelf life) or may have low bioavailability with strong side effect (Muller et al., 2004). Hence, scientific research and discoveries are going on towards development of novel drug delivery and modifying different drug delivery strategies. Drug targeting will enhance the therapeutic efficacy of drugs by reducing their side effects (Basile et al., 2012). The different strategies of drug targeting have been explained in later part. Targeted drug delivery has the ability to deliver drug in the target tissue or organ selectively and quantitatively, independent of the site and methods of administrations. The aim of targeted drug delivery is to obtain high local concentrations of drug in the target area without any side effects in normal tissues, together with low systemic exposure (Muzykantov et al., 2002; Goodman et al., 2008). Generally drug targeting has been classified into three types, which are

- A. First-order targeting—which describes delivery to a desecrate organ or tissue.
- B. Second-order targeting—it represents targeting a specific cell type (s) within the tissue or organ.
- C. Third-order targeting—it explains delivery to specific intracellular compartments in the target cells, e.g. lysosomes.

1.8   Components of Targeted Drug Delivery

1.8.1   Target

Target means specific organ or a cell or group of cells, which in chronic or acute condition need treatment by applying drug to the particular area.

1.8.2   Carrier or marker

These are required for the effective transportation of drug to the pre-selected sites. They may be called as engineered vectors, which retain drug inside or onto them either via encapsulation and/ or via spacer moiety and transport or deliver it into the target cell.

Carriers can be divided into soluble, cellular carriers, particle type.

1.8.2.1   Soluble carriers are monoclonal antibodies, modified plasma proteins, peptides

1.8.2.2   Cellular carriers are better drug carriers due to their natural biocompatibility

They will encounter endothelial barriers and can rather easily invoke an immunological response. Cellular carriers are used field cancer therapy.

1.8.2.3   Particle type carriers comprise liposomes, lipid particles, i.e. low density lipoproteins (LDL) and high density lipoproteins (HDL), polymeric micelles, nanoparticles and microspheres

1.8.3   Different pharmaceutical carriers used for drug targeting

1.8.3.1   Liposomes

Liposomes as drug delivery vehicles were first proposed by Gregoriadis, which are a novel drug delivery system (NDDS). They are vesicular structures consisting of unilamellar or multilamellar phospholipids bilayers surrounding one or several aqueous compartments. They are microscopic vesicles in which an aqueous volume is entirely enclosed by a membrane composed of lipid bilayers. Liposomes are colloidal spheres of cholesterol nontoxic surfactants, sphingolipids, glycolipids, long chain fatty acids and even membrane proteins and drug molecules or it is also called vesicular system. It differs in size, composition and charge and drug carrier loaded with variety of molecules such as small drug molecules, proteins, nucleotides or plasmids etc. A liposome consists of a region of aqueous solution inside a hydrophobic membrane. Hydrophobic chemicals can be easily dissolved into the lipid membranes, so liposomes are able to carry both hydrophilic and hydrophobic molecules (Gregoriadis et al., 1975; Gregoriadis et al., 1981; Gregoriadis et al., 1976; Gregoriadis et al., 1973).

1.8.3.2   Dendrimers

Dendrimersare the emerging new class of polymeric architectures that are which are highly branched, monodisperse macromolecules. The structure of these materials has greater impact on physical and chemical property of drug molecules. They possess empty internal cavity and many functional groups which is responsible for its high solubility. These nanostructured macromolecules have shown their potential abilities in entrapping and/ or conjugating the high molecular weight hydrophilic/hydrophobic entities by host–guest interactions and covalent bonding (prodrug approach), respectively. Despite of their extensive applications, their use in biological systems is limited due to toxicity issues associated with them (Prusty, 2012).

1.8.3.3   Monoclonal antibodies and fragments

The development of monoclonal antibodies by Kohler and Milstein in 1975 brought a new insight using antibody therapy for disease. Since from their discovery, the number of preclinical and clinical studies using monoclonal antibodies has greatly increased for the treatment of the disease. They are used for recognition antibodies for cancer therapy. The advent of monoclonal antibodies (MoAbs) has already made an impact on cancer diagnosis, particularly in the areas of immuno scintigraphy and immuno histology. The discovery of recombinant DNA technology led to the development of antibodies and fragments that are tailored for the optimal behaviour *in vivo* (Gaetana et al., 2014; Jain, 2014).

1.8.3.4   Microspheres and nanoparticles

Microspheres and nanoparticles often consist of biocompatible polymers and belong either to the soluble or the particle type carriers. Microspheres can be defined as structure made up of continuous phase of one or more miscible polymers in which drug particles are dispersed at the molecular or macroscopic level. It has a particle size of (1–1000 nm). Nanoparticles are defined as particulate dispersions or solid particles with a size in the range of 10–1000 nm. The drug is dissolved, entrapped, encapsulated or attached to a nanoparticle matrix. Nanoparticles are smaller (0.2–0.5µm) than microspheres (30–200 µm) and may have a smaller drug loading capacity than the soluble polymers. After systemic administration, these drug delivery carriers are quickly distributed. Besides parenteral application of microspheres and nanoparticles for cell selective delivery of drugs, they have more recently been studied for their application in oral delivery of peptides and peptidomimetics, which are small protein-like chain designed to mimic a peptide (Shive et al., 1997).

1.8.3.5   Polymeric micelles

Polymeric micelles are characterized by a core–shell structure. They have a di-block structure with a hydrophilic shell and a hydrophobic core. The hydrophobic core generally consists of a biodegradable polymer that serves as a reservoir for an insoluble drug. Non- or poorly biodegradable polymers can be used, as long as they are not toxic to cells and can be renally secreted. If a water-soluble polymeric core is used, it is rendered hydrophobic by chemical conjugation with a hydrophobic drug. The viscosity of the micellar core may influence the physical stability of the micelles as well as drug release. The bio-distribution of the micelle is mainly dictated by the nature of the shell which is also responsible for micelle stabilization and interactions with plasma proteins and cell membranes. The micelles can contain functional groups at their surface for conjugation with a targeting moiety. Polymeric micelles are mostly small (10–100 nm) in size and drugs can be incorporated by chemical conjugation or physical entrapment. For efficient delivery activity, they should maintain their integrity for a sufficient amount of time after injection into the body. Most of the experience with polymeric micelles has been obtained in the field of passive targeting of anticancer drugs to tumours (Croy et al., 2006).

1.9   Strategies of Drug Targeting

Different strategies are used for drug targeting to the desired organ/tissue as shown in below Figure 1.1. These strategies are as follows:

Figure 1.1: Different strategies for drug targeting

1.9.1   Passive targeting

Drug delivery systems which are targeted to systemic circulation are characterized as passive delivery systems. In this technique drug targeting occurs because of the body's natural response to physicochemical characteristics of the drug or drug carrier system. The ability of some colloid to be taken up by the reticuloendothelial systems (RES) especially in liver and spleen made them ideal substrate for passive hepatic targeting of drugs.

1.9.2   Inverse targeting

As in this type of targeting passive uptake of colloidal carrier by RES is avoided and hence the process is referred to as inverse targeting. To achieve inverse targeting, RES normal function is suppressed by pre injecting large amount of blank colloidal carriers or macromolecules like dextran sulphate. This approach leads to saturation of RES and suppression of defence mechanism. This type of targeting is an effective approach to target drug(s) to non-RES organs.

1.9.3   Active targeting

In this approach, carrier system bearing drug reaches to specific site on the basis of modification made on its surface rather than natural uptake by RES. Surface modification technique, include coating of surface with either a bioadhesive, non-ionic surfactant or specific cell or tissue antibodies (i.e. monoclonal antibodies) or by albumin protein.

1.9.4   Ligand-mediated targeting

Achieved using specific mechanisms such as receptor dependent uptake of natural LDL particles and synthetic lipid micro emulsions of partially reconstituted LDL particles coated with the apoproteins.

1.9.5   Physical targeting

In this type of targeting some characteristics of environment changes like pH, temperature, light intensity, electric field, and ionic strength small and even specific stimuli like glucose concentration are used to localize the drug carrier to predetermined site. This approach was found exceptional for tumour targeting as well as cytosolic delivery of entrapped drug or genetic material.

1.9.6   Dual targeting

In this targeting approach, carrier molecule itself has their own therapeutic activity and thus increasing the therapeutic effect of drug. For example, a carrier molecule having its own antiviral activity can be loaded with antiviral drug and the net synergistic effect of drug conjugate was observed.

1.9.7   Double targeting

When temporal and spatial methodologies are combined to target a carrier system, then targeting may be called double targeting. Spatial placement relates to targeting drugs to specific organs tissues, cells or even subs cellular compartment. Whereas temporal delivery refers to controlling the rate of drug delivery to target site.

1.10   Biological Processes and Events Involved in Drug Targeting

1.10.1   Cellular uptake and processing

The major objectives for targeted drug delivery are enhancing drug accumulation at the target site by reducing the nondiscriminate uptake of toxic agents. In order to target drugs to specific tissue systems within the body, drug molecules can be directly attached to a targeting agent or complexed with a vehicle, or macromolecule, that contains targeting moieties. Macromolecules can be bioengineered to incorporate a variety of synthetic and natural compounds, including drugs, ligands, and radionuclide. Following the administration lower molar mass drugs can enter into or pass through various cells by simple diffusion process. Targeted drug delivery usually has macromolecular assemblies, so enter by a process called endocytosis. Endocytosis is the process by which the materials move into the cell. There are three types of endocytosis, i.e. phagocytosis, pinocytosis, and receptor-mediated endocytosis as shown in Table 1.1. In phagocytosis or "cellular eating," the cell's plasma membrane surrounds or engulfs a macromolecule (drug) or even an entire cell from the extracellular environment and buds off to form a food vacuole or phagosome by opsonization process. The newly formed phagosome then fuses with a lysosome whose hydrolytic enzymes digest the "food" inside (Marsh Mark, 2001).The phagocytic pathway as shown in Figure 1.2 consists of three distinct steps:

1. Recognition of the particles by opsonization in the bloodstream.
2. Adhesion of the opsonized particles onto the cell membrane.
3. Ingestion of the particle by the cells.

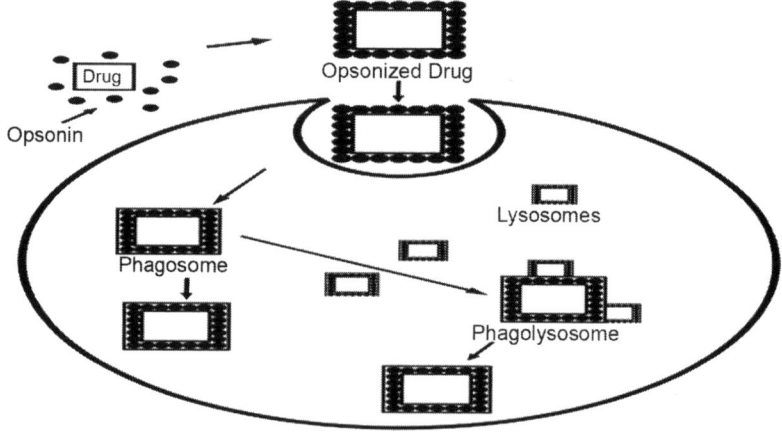

Figure 1.2: Drug diffusion by phagocytic mechanism

Table 1.1: Classification of different drug targeting pathways

Specific pathway	Pathway	Definition
Endocytosis	Clathrin- and caveolin-mediated	Energy-dependent process by which cells internalize biomolecules
Phagocytosis	Mannose receptor-, complement receptor- and scavenger receptor-mediated	Actin-dependent endocytic process by which professional phagocytes (macrophages, dendritic cells and neutrophils) engulf particles with sizes larger than 0.5 μm
Pinocytosis		Endocytic process by which cells absorb extracellular fluids, small molecules and small vesicles (~100 nm) are formed

However, in pinocytosis or "cellular drinking," the cell engulfs drops of fluid by pinching in and forming vesicles that are smaller than the phagosomes formed in phagocytosis. Like phagocytosis, pinocytosis is a nonspecific process in which the cell takes in whatever solutes that are dissolved in the liquid it envelops. Pinocytosis is divided into two types: first is fluid phases pinocytosis and other is adsorptive pinocytosis (Marsh Mark, 2001). Diagrammatic representation of pinocytosis is shown in Figure 1.3.

Pinocytosis

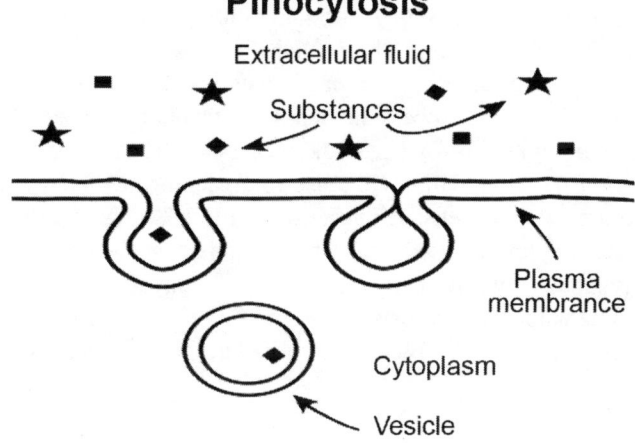

Figure 1.3: Fluid uptake by pinocytic mechanism

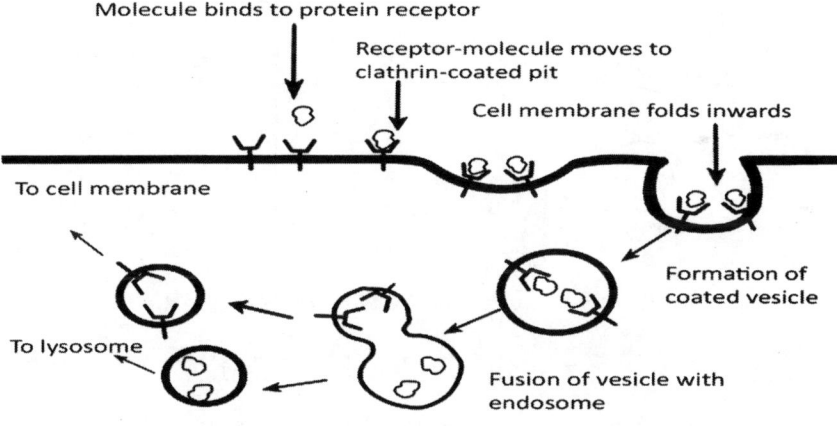

Figure 1.4: Receptor-mediated endocytosis mechanism

Unlike phagocytosis and pinocytosis, receptor-mediated endocytosis as shown in Figure 1.4 is an extremely selective process of importing materials into the cell. This specificity is mediated by receptor proteins located on depressed areas of the cell membrane called coated pits. Macromolecules, such as sugars and hormones bind with the receptor protein in cell membrane and this binding not only form invagination but also form a protein coat around vesicles called clathrin coated vesicles. The cytosolic surface of coated pits is covered by coat proteins. In receptor-mediated endocytosis, the cell will only take in an extracellular molecule if it binds to its specific receptor protein on the cell's surface. Once bound, the coated pit on which the bound receptor protein is located then invaginates, or pinches in, to form a coated vesicle. Similar to the digestive process in non-specific phagocytosis, this coated vesicle then fuses with a lysosome to digest the engulfed material and release it into the cytosol. Mammalian cells use the receptor-mediated endocytosis to take cholesterol into cells. Cholesterol in the blood is usually found in lipid–protein complexes called low-density lipoproteins (LDLs) which bind to specific receptor proteins on the cell surface, thereby triggering their uptake by receptor-mediated endocytosis (McMahon et al., 2011). A diagrammatic representation of different types of endocytic phenomena is shown in Figure 1.5.

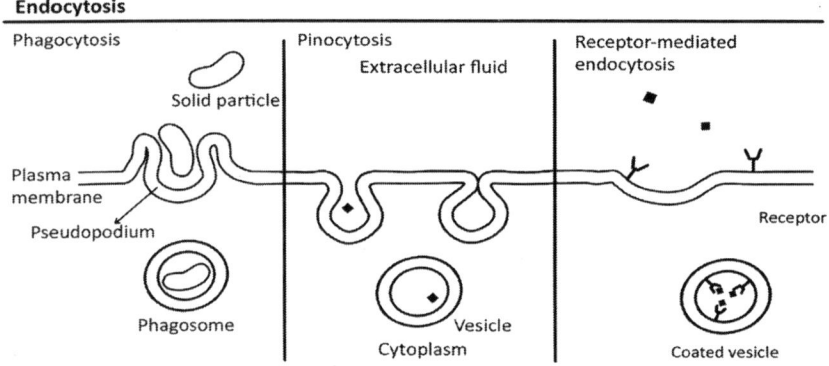

Figure 1.5: Different types of endocytic mechanism

1.10.2   Transport across the epithelial barrier

The oral, buccal, nasal, vaginal, and rectal cavities are internally lined with one or more layers of epithelial cells. Low molar mass drugs cross the above by passive diffusion, carrier-mediated transport. Passive transport is usually higher in damaged mucosa where as active transport depends on structural integrity of epithelial cells. Positively charged particles showed increased uptake than negatively charged counter parts (Christian Tscheik et al., 2013).

1.10.3   Extravasations

Many diseases result from the dysfunction of cells located outside the cardiovascular system. Thus for a drug to exert its therapeutic effects it must exit from the central circulation and this process of transvascular exchange is called extravasations which is governed by blood capillary walls. Generally control permeability of capillaries depends on structure of the capillary wall, pathological condition, rate of blood and lymph supply and different physicochemical factors of drug. The structure of the blood capillary varies in different organs tissues. It consists of a single layer of endothelial cells joined together by intercellular junctions. Depending on the morphology and continuity of the endothelial layer and the basement membrane blood capillaries are divided into three types, i.e. continuous, fenestrated and sinusoidal. Continuous capillaries are common and widely distributed in the body exhibit tight inter endothelial junctions and an uninterrupted basement membrane. Similarly fenestrated capillaries shows interendothelial gaps of 20–80 nm and sinusoidal capillaries show inter endothelial gaps of 150 nm. Macromolecules can transverse the normal endothelium by passive process, such as nonspecific fluid phase trans capillary pinocytosis and passage through inter endothelial junctions gaps or fenestrate or by receptor-mediated transport systems. Organs, such as the lung with very large surface areas have a proportionately large total permeability and consequently high extravasations. The endothelium of brain is the strongest of all endothelia formed by continuous non fenestrated endothelial cells, which show no pinocytic activity. Soluble macromolecules permeate the endothelial barrier more readily than particulate macromolecules by which the rate of movement of fluid across the endothelium appears to be directly related to the difference between the hydrostatic and osmotic forces (Christian Tscheik et al., 2013).

1.10.4   Lymphatic uptake

Following extravasations, the drug molecules can either reabsorb into the blood stream directly by the enlarged post capillary interendothelial cell pores found in most tissues or enter into the lymphatic system and then return with the lymph to the blood circulation. Drugs administered through subcutaneous, intramuscular, transdermal and peritoneal routes reach the systemic circulation by lymphatic system. There are some advantages of lymphatic absorption of drugs as it avoids first pass effect. Compounds of high molecular weight (above 16,000) can be absorbed by lymphatic transport. Targeted delivery of drugs to lymphatic system as in certain case of cancer is also possible. The direct delivery of drugs into lymphatic has been proposed as a potential approach to kill malignant lymphoid cells located in lymph nodes. Although lymph flow is exceptionally slow, but fats, fat-soluble vitamins and highly lipophilic drugs are absorbed through lymphatic circulation (Christian Tscheik et al., 2013).

1.11   Tumour Targeting and Brain-Specific Drug Delivery

A tumour is defined as a swelling or morbid enlargement that result from an overabundance of cell growth and division. The principal barriers associated with tumour targeting comprise peculiar tumour vasculature that principally comprises heterogeneous blood flow and vascular resistance. This heterogeneity leads to uneven distribution of administered therapeutics which often leading to poor therapeutic response. The principal goals of the targeted drug delivery system is aimed at protection of the drug in concern to the site of action from the metabolic degradation or inactivation during transit, particularity for specific target devoid of any nonspecific interactions with the host tissues and penetration of relevant concentrations of drug within the tumour tissues for therapeutic responses. Drug targeting strategies have frequently been divided into categories of "passive" and "active."

In passive targeting drug accumulation occurs in the areas around the tumours with leaky vasculature which referred as the enhanced permeation and retention (EPR) effect. The natural bio-distribution pattern of the drug delivery carrier is exploited for its preferential localization in the vicinity of the tumours such as enhanced permeation and retention effects, phagocytosis of particulate carrier by mononuclear phagocytises systems (MPS) and preferential localization in the organs of reticuloendothelial system (RES). In addition, other typical properties of the tumour microenvironment such as low extracellular pH, relative micro-acidosis, mild hyperthermia, etc. could also be employed for availing passive targeting of therapeutics.

Active targeting describes specific interactions between drug/drug carrier and the target cells through specific ligand–receptor interactions. Ligand–receptor interactions occur only when the two components are in close proximity. In active targeting "ligand–receptor type interaction" for intracellular localization occurs after blood circulation and extravasations. This is why increasing blood circulation time by PEGylation (i.e. modifying the surface of nanoparticles with polyethylene glycol) and/or improving the EPR effect is expected to enhance delivery to the tumour site (Beduneau et al., 2007;Deckert, 2009; Hong et al., 2009;Zensi et al., 2009; Canal et al., 2010).

The brain is a delicate organ, and evolution built very efficient ways to protect it. During various neurological disorders, the drug delivery to brain requires careful application. The delivery of drugs to central nervous system (CNS) is a challenge in the treatment of neurological disorders. Drugs may be administered directly into the CNS or administered systematically (e.g. by intravenous injection) for targeted action in the CNS. The major challenge to CNS drug delivery is the blood–brain barrier (BBB), which limits the access of drugs to the brain substance. Advances in understanding of the cell biology of the BBB have opened new avenues and possibilities for improved drug delivery to the CNS. Brain capillaries are lined with a layer of special endothelial cells that lack fenestrations and are sealed with tight junctions called zona occludens. Various transmembrane proteins such as occludin and claudin that project into and seal the paracellular pathway form the tight junctions in brain capillaries. The interaction of these junctional proteins is complex and effectively blocks an aqueous route of free diffusion for polar solutes from blood along with solutes free access to brain interstitial (extracellular) fluid. Drugs, that are effective against CNS diseases and reach the brain via the blood compartment, must pass through the BBB. The BBB is mainly formed by brain capillary endothelial cells (BCEC), although other cell types, such as astrocytes form the structural framework for the neurons and control their biochemical environment. Oligodendrocytes are responsible for the formation and maintenance of the myelin sheath, which surround the axons and is essential for the fast transmission of action potential by salutatory conduction. To bypass the BBB and to deliver therapeutics into the brain, three different approaches are currently used. All these approaches are relatively costly, require anaesthesia and hospitalization and are non-patient friendly. These techniques may enhance tumour dissemination after successful disruption of the BBB. Neurons may be damaged permanently from unwanted blood components entering in to the brain. For delivery of drug to brain by crossing blood–brain barrier requires osmotic and chemical opening of the blood–brain barrier as well as the use of transport or carrier systems. Sometimes various pharmacological agents like dimethyl sulfoxide (DMSO) have been used to open the BBB and direct invasive methods can introduce therapeutic agents into the brain substance. It is important to consider not only the net delivery of the agent to the CNS, but also the ability of the agent to access the relevant target site within the CNS. Targeting the brain the blood–brain barrier (BBB) is a unique protective barrier that provides an efficient exclusion by obstructing the free flow of blood between the brain and the rest of the body. This also prevents penetration of hydrophilic compounds such as various neurotransmitters, amino acids etc. unless these are transported to the brain by an active transport system. Hence, the drug carrier complex must be sufficiently lipophilic for distribution throughout the body after intravenous administration (Bhupinder Singh et al., 2011).

1.11.1   Different strategies for drug targeting to brain

1.11.1.1   *Lipid-mediated transport*

The penetration of a xenobiotic into the CNS tissue is due to its low molecular weight, lack of ionization at physiological pH and lipophilicity. Despite extensive applications of medicinal chemistry, till date, there is not even a single FDA approved drug that exemplifies the conversion of a poor brain-penetrating molecule into a high brain penetrating one. Hence, scientist are trying for lipidization of water soluble drugs, whereby chemical medications are used to block existing hydrogen bond forming groups on the parent drug molecule (Dwibhashyam et al., 2008).

1.11.1.2   Carrier-mediated transport

An alternative technique to increase brain penetration is to modify drugs, such that there is increased carrier-mediation of the drug. For instance, a-carboxylation of the water-soluble catecholamine drug results in the formation of a neutral amino acid. Although the BBB penetration of the catecholamine is very low, but the amino acid may then penetrate the BBB at pharmacologically, significant rates via carrier-mediated transport (Huwyler et al., 1996).

1.11.1.3   Intraventricular or intrathecal route

The intra cerebro ventricular (ICV) approach injects drug into the cerebrospinal fluid (CSF) compartment, which is reported to be 140 ml in volume in humans. The entire CSF pool in the human brain is turned over every 4–5 h and four to five times per day. Drugs can be infused intraventricularly using an Ommaya reservoir, a plastic basin implanted subcutaneously in the scalp and connected to the ventricles within the brain via an outlet catheter (Huwyler et al., 1996).

1.11.1.4   Intranasal drug administration

The neural connections between the nasal mucosa and the brain provide a unique pathway for the delivery of therapeutic agents to the CNS. Intranasally administered therapeutic agents reach the CNS via the olfactory and trigeminal neural pathways. More recently, the contribution made by the trigeminal pathway to intranasal delivery to the CNS has also been recognized, especially to caudal brain regions and the spinal cord. Extracellular delivery, rather than axonal transport, is strongly indicated by the short time frame (=10 minutes) observed for intranasal therapeutics to reach the brain from the nasal mucosa. Possible mechanisms of transport may involve bulk flow and diffusion within perineuronal channels, perivascular spaces, or lymphatic channels directly connected to brain tissue or cerebrospinal fluid. An obvious advantage of this method vis-à-vis other strategies is that it is noninvasive. Nonetheless, there have been certain difficulties that need to be overcome to achieve successful brain drug delivery through nasal route, like an enzymatically active and low pH nasal epithelium, and possibility of mucosal irritation, or the possibility of large variability caused by nasal pathology, such as common cold (Huwyler et al., 1996).

1.11.1.5   Brain drug targeting through liposome

These liposomal carriers access the brain from blood via receptor-mediated transcytosis and deliver their content (small drug molecules, plasmid) into the brain parenchyma, without damaging the BBB. This requires the presence of receptor-specific targeting ligands at the tip of 1–2% of the PEG 2000 strands. Targeting ligands are peptide mimetic monoclonal antibodies, i.e. able to trigger the activation of receptors (transferring or insulin receptors) that are highly expressed on the brain capillary endothelium. These antibodies, directed against external receptor epitopes, do not interfere with the natural ligand binding sites, thus avoiding competition. Colloidal carriers should have diameter less than 100 nm to fit the loading capacity of these transport systems (Kreuter et al., 1995).

1.11.1.6   Nanoparticulate drug delivery

Nanoparticles are solid colloidal particles, ranging in size from 1 to 1000 nm, and consisting of various macromolecules in which drugs could be adsorbed, entrapped, or covalently attached. Generally administered by the intravenous route like liposomes, they have been developed for the targeted delivery of therapeutic or imaging agents. Their stellar merits over liposomes comprise, low number of excipients used in their formulations, the simple procedures for preparation, a high physical stability, and the possibility of sustained drug release that may be suitable in the treatment of chronic diseases. Strategies for nanoparticulates targeting to the brain rely on their interaction with specific receptor-mediated transport systems in the BBB (Dwibhashyam et al., 2008).

1.11.1.7   Biochemical blood–brain disruption

Biochemical disruption of BBB is now a days a safer technique for delivery of drug to brain. Selective openings of brain tumour capillaries by intracarotid infusion of leukotriene C4 is achieved without concomitant alteration of the adjacent BBB. The biochemical opening utilizes a novel observation that normal brain capillaries appear to be unaffected when vasoactive leukotriene treatments are used to increase their permeability (Dwibhashyam et al., 2008).

1.11.1.8   Prodrug with improved brain permeability

Brain uptake of drugs can be efficiently ameliorated via prodrug formation. Prodrugs are pharmacologically inactive compounds that result from transient chemical modifications of biologically active species. The chemical change is usually designed to improve membrane permeability or water solubility.

Following administration, the prodrug converted to the active form, is brought closer to the receptor site, and is maintained there for longer periods of time (Dwibhashyam et al., 2008).

1.11.1.9   Chemical delivery systems

In involves a complex formation between drug and the lipophilic dihydropyridine carrier, which after systemic administration readily transverses the BBB because of its lipophilicity. After entering in to brain, the dihydropyridine moiety is enzymatically oxidized to the charged ionic pyridinium salt, which has the dual effect of accelerating the rate of systemic elimination by the kidney and bile and trapping the drug–pyridinium salt complex inside the brain. Biotechnological approaches recombinant protein neurotherapeutics can be delivered across the human BBB following genetic engineering, expression, and purification of recombinant fusion proteins. In this approach, the nontransportable protein therapeutic, e.g. a neurotrophin, is fused to the carboxyl or aminoterminus of either the heavy chain or light chain of the genetically engineered monoclonal antibody, thus enabling the complex to cross BBB (Dwibhashyam et al., 2008).

1.12   Conclusions

The drug targeting is to achieve a desired pharmacological response at desired site without any undesirable side effects and interactions at other sites. As we came to know drug targeting can be achieved by any two approaches, i.e. eider by doing chemical modification of a parent compound to a derivative which is activated only at the target site or by utilizing carriers system such as liposomes, microspheres, nanoparticles and macromolecules to direct the drug to its site of action. Drug targeting is an effective approach in avoidance of hepatic first pass metabolism, rapid onset of action, better patient compliance, enhancement of bioavailability etc. Hence, there is a need to develop novel drug delivery systems to achieve better drug targeting by this approach.

1.13   References

Beduneau, A., Saulnier, P., Hindre, F., Clavreul, A., Leroux, J.C. and Benoit, J.P. (2007), 'Design of targeted lipid nanocapsules by conjugation of whole antibodies and antibody Fab' fragments', Biomaterials, 28, 4978–4990.

Basile, L., Pignatello, R. and Passirani, C. (2012), 'Active targeting strategies for anticancer drug nanocarriers', Curr. Drug Deliv. 9(3), 255–268.

Bhupinder, S., and Rishi, K. (2011), 'Brain drug delivery: Problems and prospects', Chron. Pharm., 63rd Indian Pharmaceutical Congress, A Special Supplement, 26–27.

Brahmankar, D.M., and Jaiswal, S.B. (2009), Biopharmaceutics and Pharmacokinetics. 2nd Edition, Delhi, Vallabh Prakashan, pp. 10–17.

Bhargav, E., Madhuri, N., Ramesh, K. and Anandmanne, R.V. (2013), 'Targetted drug delivery-A review', World J. Pharm. Pharmaceut. Sci., 3, 150–169.

Canal, F., Vicent, M.J., Pasut, G. and Schiavon, O. (2010), 'Relevance of folic acid/polymer ratio in targeted PEG–epirubicin conjugates', J. Control. Release, 146, 388–399.

Christian, T., Ingolf, E.L. and W. (2013), 'Trends in drug delivery through tissue barriers containing tight junctions', Tissue Barr., 1(2), e24565-1–e24565-8.

Croy, S.R., and Kwon, G.S. (2006), 'Polymeric micelles for drug delivery', Curr. Pharm. Des., 12, 4669–4684.

Dwibhashyam, V.S.N.M., and Nagappa, A.N. (2008), 'Strategies for enhanced drug delivery to the central nervous system', Indian J. Pharm. Sci., 70, 145–153.

Ehrlich, P. (1960), The collected papers of Paul Ehrlich, London, Pergamon, pp. 1854–1915.

Gerber, D.E. (2008), 'Targeted therapies: A new generation of cancer treatments', Am. Fam. Phys., 77, 311–319.

Gaetana D.F., et al. (2011), 'Monoclonal antibodies and antibody fragments: state of the art and future perspectives in the treatment of nonhaematological tumors', Expert Opin Biol Ther. 11, 1433–1445.

Goodman, T.T., Chen, J., Matveev, K. and Pun, S.H. (2008), 'Spatio-temporal modeling of nanoparticle delivery to multicellular tumor spheroids', Biotechnol. Bioeng., 101(2), 388–399.

Gregoriadis, G., and Neerunjun, D.E. (1975), 'Homing of Liposomes to target cells', Biochem. Biophys. Res. Commun., 65, 537–544.

Gregoriadis, G. (1981), 'Targeting of drug: implications in medicine'. Lancet, 2, 241–247.

Gregoriadis, G. (1976), 'The carrier potential of liposomes in biology and medicine (first of two parts).' N. Engl. J. Med., 295(13), 704–710.

Gregoriadis, G. (1973), 'Drug entrapment in liposomes', FEBS Lett., 36(3), 292–296.

Gupta, M., and Sharma, V. (2011), 'Targeted drug delivery system: A review,' Res. J. Chem. Sci., 1, 134–138.

Hong, M., Zhu, S., Jiang, Y., Tang, G. and Pei, Y. (2009), 'Efficient tumor targeting of hydroxycamptothecin loaded PEGylated niosomes modified with transferrin', J. Control. Release, 133, 96–102.

Huwyler, J., Wu, D., and Pardridge, W.M. (1996), 'Brain drug delivery of small molecules using immunoliposomes', Proc. Natl. Acad. Sci. USA, 93, 14164–14169.

Kreuter, J., Alyautdin, R.N., Kharkevich, D.A. and Ivanor, A.A. (1995), 'Passage of peptides through the blood–Brain barrier with colloidal polymer particles nanoparticles', Brain Res., 674, 171–174.

Mimeault, M., Hauke, R. and Batra, S.K. (2008), 'Recent advances on the molecular mechanisms involved in the drug resistance of cancer cells and novel targeting therapies', Clin. Pharmacol. Ther., 83, 673–691.

Mills, J.K., and Needham, D. (1999), 'Targeted drug delivery', Expert Opin. Ther., 9, 1499–1513.

Muller, R.H., and Keck, C.M. (2004), 'Challenges and solutions for the delivery of biotech drugs-A review of drug nanocrystal technology and lipid nanoparticles', J. Biotechnol., 113, 151–170.

Muzykantov, V., and Torchilin, V. (2002), 'Biomedical Aspects of Drug Targeting', Springer Science & Business Media, LLC, New York, pp. 3–44.

Marsh Mark. (2001), Endocytosis. New York, Oxford University, pp. 1–25.

McMahon, H.T., and Boucrot, E. (2011), Molecular mechanism and physiological functions of clathrin-mediated endocytosis. Nat. Rev. Mol. Cell Biol., 12, 517.

Jain, N.K. (2014). Progress in Controlled and Novel Drug Delivery Systems, New Delhi, CBS Publishers & Distributors Pvt. Ltd.

Prusty A. (2012), 'Dendrimer, The recent drug delivery system', Int. Res. J. Pharm., 3, 10–12.

Deckert, P.M. (2009), 'Current constructs and targets in clinical development for antibody based cancer therapy', Curr. Drug Targ., 10, 158–175.

Shive, M.S., and Anderson, J.M. (1997), Biodegradation and biocompatibility of PLA and PLGA microspheres Adv. Drug Deliv. Rev., 28, 5–24.

Strebhardt, K., and Ullrich, A. (2008), 'Paul Ehrlich's magic bullet concept: 100 years of progress', Nat. Rev. Cancer, 8, 473–480.

Torchilin, V.P. (2000), 'Drug targeting', Eur. J. Pharm. Sci., 11, S81–S91.

Vyas, S.P., Khar, R.K., (2002), Targeted & Controlled Drug Delivery: Novel Carrier Systems, New Delhi, CBS Publishers.

Zensi, A., et al. (2009), 'Albumin nanoparticles targeted with ApoE enter the CNS by transcytosis and are delivered to neurons', J. Control. Release, 137, 78–86.

List of Abbreviations

BBB	:	Blood–Brain Barrier
BCEC	:	Brain Capillary Endothelial Cells
CNS	:	Central Nervous System
DMSO	:	Dimethyl Sulfoxide
EPR	:	Enhanced Permeation and Retention
HDL	:	High Density Lipoproteins
LDL	:	Low Density Lipoproteins
MoAbs	:	Monoclonal Antibodies
NDDS	:	Novel Drug Delivery System
nm	:	Nanometer
RES	:	Reticuloendothelial systems

2

Targeted Methods: Nanoparticles and Liposomes: Types, Preparation and Evaluation

Kahnu Charan Panigrahi[*1], Chinam Niranjan Patra[1], Debashish Ghose[1], Suryakanta Swain[2] and Bikash Ranjan Jena[2]

[1] Department of Pharmaceutics, Roland Institute of Pharmaceutical Sciences, Berhampur-760010, Odisha, India.
[2] Southern Institute of Medical Sciences, College of Pharmacy, Department of Pharmaceutics, Guntur-522001, Andhra Pradesh, India.

2.1 Introduction

The research and developments at atomic, molecular, and macromolecular scales, which lead to the controlled manipulation and study of structures in the range of 1–100 nm had given rise to world of nanotechnology. The nanoparticles fit into colloidal drug delivery systems, which offer advantages of drug targeting by modified body distribution as well as the improvement of the cellular uptake, which contributes from reduction of undesired toxic side effects of the free drugs. Albumin was first incorporated to be designed in the form biopolymeric nanoparticles and nonbiodegradable synthetic polymers like poly (methylacrylate) and polyacrylamide. Chronic toxicity due to the tissue overloading of nondegradable polymers was considered as major setback for the systemic administration of polyacrylamides and poly (methylacrylate) nanoparticles in humans. As a result,the nanoparticles received much attention which was designed with synthetic biodegradable polymers including polyalkylcyanoacrylate, poly (lactic-*co*-glycolic acid) and polyanhydride (Verdun et al., 1990; Quintanar-Guerrero et al., 1998). PLGA poly(lactic-*co*-glycolic acid) has gained great acceptability in the medical and pharmaceutical field because of its biodegradability, toxicological safety and desirable biocompatibility in the last 10 years. Nowadays, PLGA is widely applied in controlled drug delivery systems, including nano- and microparticles and implants (Klose et al., 2010).Across the globe, numerous scientists have developed various methods, such as solvent evaporation, nanoprecipitation, salting-out, emulsification–diffusion and supercritical fluid technology for preparation of nanoparticles (Soppimath et al., 2001).

Since the discovery of liposomes by Bangham and Hornein (1964), the potential of liposomes as drug delivery carriers has been extensively explored via versatile administrative routes, such as parenteral, oral, pulmonary, nasal, ocular and transdermal (Bangham and Horne, 1964; Nekkanti and Kalepu,

2015). Liposomes abundance utilizations for therapeutic benefit have been continuously expanded due to their flexible structures and practical function since their introduction (Balazs and Godbey, 2011; Zylberberg and Matosevic, 2016). Among the variety of liposome modifications, active targeting emerges as a useful and common strategy to enhance liposomal delivery system's desired properties, such as targeting ability, increased selectivity, increased cellular internalization, prolonged duration of exposure, minimized adverse effects, and improved therapeutic index (Puri et al., 2009; Bardania, Tarvirdipour and Dorkoosh, 2017).

2.2   Classification of Nanoparticles

According to the structural organization, biodegradable polymeric nanoparticles are classified as nanocapsules and nanospheres. The drug molecules are either entrapped inside or adsorbed on the surface.

2.2.1   Nanocapsules

Nanocapsules are spherical hollow structures in which the drug is confined in the cavity and surrounded by a polymer membrane. Size between 50 and 300 nm is preferred for drug delivery and they must be filled with oil, which can dissolve lipophilic drugs.

2.2.2   Nanospheres

A polymeric nanosphere is defined as matrix type, solid colloidal particle in which drugs are dissolved, entrapped, encapsulated, chemically bound or absorbed to the constituent polymer matrix. These particles are typically larger than micelles having the diameters between 100 and 200 nm and may also display considerably more polydispersity.

2.2.3   Solid lipid nanoparticles (SLNs)

Solid lipid nanoparticles (SLNs) are submicron sized particles comprises biocompatible and biodegradable materials, such as triglycerides and fatty acids. They offer a prominent advantage over the NPS as they are made of physiological lipids and surfactants which are recognized as safe.

2.2.4   Polymeric nanoparticles

Polymeric (natural or synthetic) nanoparticles are particles of diameter below 1 μm. The use of natural polymers (i.e. proteins or polysaccharides) is restricted, due to the variation in purity, requirement of crosslinking and denaturation of the drug. The most widely used synthetic polymers are PLA, PLGA and PACA, respectively.

2.3   Production of Drug Nanoparticles

Drug-loaded nanoparticles can be developed different approaches of technologies such as; "bottom-up" and "top-down approach." Controlled crystallization/precipitation by adding a suitable nonsolvent are major bottom-up methods. Milling or homogenization are associated with top-up technology. However, techniques that involve pretreatment step followed by reduction in particle size are also being used to produce nanoparticles with the desired size distribution. Dispersion of drug in preformed polymers is a common technique used to prepare biodegradable nanoparticles from poly (lactic acid) (PLA), poly (d,l-glycolide) (PLGA), poly (d, l-lactide-*co*-glycolide) and poly (cyanoacrylate) (PCA). These can be accomplished by different methods described below:

2.3.1   Solvent evaporation

In this method, the suitable polymer is dissolved in an organic solvent, such as dichloromethane, chloroform or ethyl acetate which is also used as the solvent for dissolving the hydrophobic drug, mentioned in Figure 2.1.The mixture of polymer and drug solution is then emulsified in an aqueous solution containing a surfactant or emulsifying agent to form an oil in water (o/w) emulsion. After the formation of stable emulsion, the organic solvent is evaporated either by reducing the pressure or by continuous stirring (Pinto Reis et al., 2006).

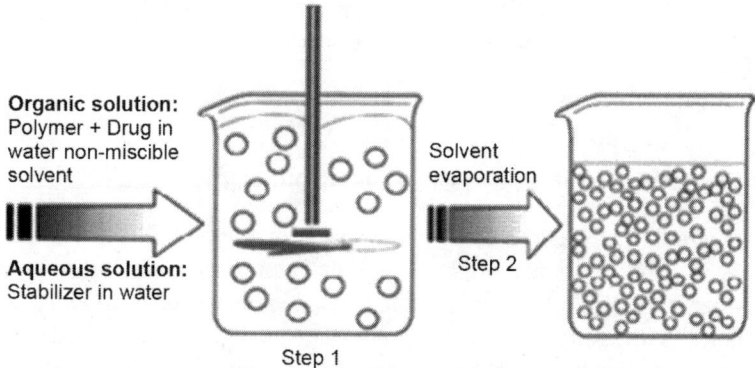

Figure 2.1: Schematic representation of the solvent–evaporation technique

2.3.2   Nanoprecipitation

Nanoprecipitation is also called solvent displacement method. It involves the precipitation of a preformed polymer from an organic solution and the diffusion of the organic solvent in the aqueous medium in the presence or absence of a surfactant (Quintanar-Guerrero et al., 1998; Barichello et al., 1999).The polymer is dissolved in a water-miscible solvent of intermediate polarity, leading to the precipitation of nanospheres. This phase is injected into a stirred aqueous solution containing a stabilizer as a surfactant, mentioned in Figure 2.2. Polymer deposition on the interface between the water and the organic solvent, caused by fast diffusion of the solvent, leads to the instantaneous formation of a colloidal suspension.

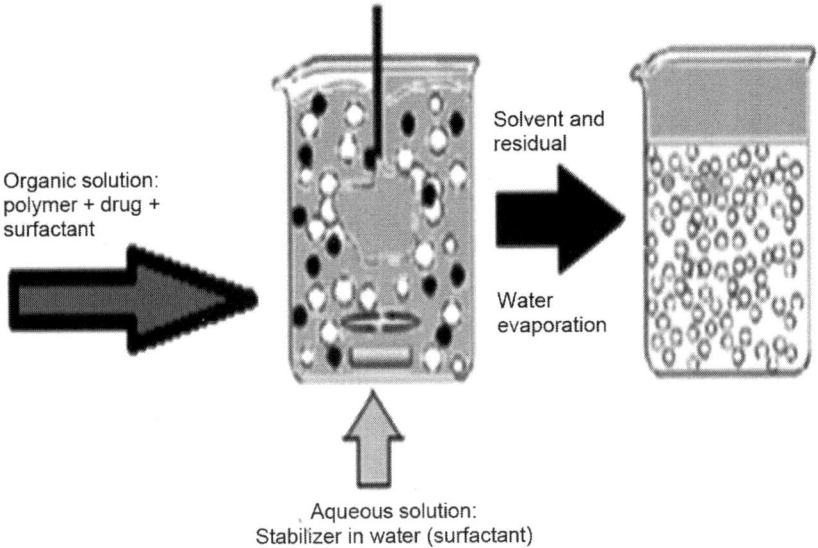

Organic solution: polymer + drug + surfactant

Solvent and residual

Water evaporation

Aqueous solution: Stabilizer in water (surfactant)

Figure 2.2: Schematic representation of the nanoprecipitation technique

2.3.3   Emulsification or solvent diffusion

This is a modified version of solvent evaporation method (Niwa et al., 1993). In this method, the water miscible solvent along with a small amount of the water immiscible organic solvent is used as an oil phase. Owing to the spontaneous diffusion of solvents, an interfacial turbulence is created between the two phases leading to the formation of small particles.

2.3.4   Salting-out

Salting-out is based on the separation of a water miscible solvent from aqueous solution via a salting-out effect. The salting-out procedure can be considered as a modification of the emulsification/solvent diffusion. Polymer and drug are initially dissolved in a solvent, such as acetone, which is subsequently emulsified into an aqueous gel containing the salting-out agent (electrolytes, such as magnesium chloride, calcium chloride, and magnesium acetate, or nonelectrolytes, such as sucrose) and a colloidal stabilizer, such as polyvinylpyrrolidone or hydroxyethyl cellulose. This oil/water emulsion is diluted with a sufficient volume of water or aqueous solution to enhance the diffusion of acetone into the aqueous phase; thus, inducing the formation of nanospheres (Pinto Reis et al., 2006).

2.3.5   Dialysis

Dialysis offers a simple and effective method for the preparation of small, narrow-distributed (Jeon et al., 2000). Polymer is dissolved in an organic solvent and placed inside a dialysis tube with proper molecular weight cutoff. Dialysis is performed against a nonsolvent miscible with the former miscible.

2.3.6   Supercritical fluid technology (SCF)

The need to develop environmentally safer methods for the production of PNP has motivated research on the utility of supercritical fluids as more environmental friendly solvents, with the potential to produce PNPs with high purity and without any trace of organic solvent (Niwa et al., 1993; York, 1999). Two principles have been developed for the production of nanoparticles using supercritical fluids:

1. Rapid expansion of supercritical solution (RESS)
2. Rapid expansion of supercritical solution into liquid solvent (RESOLV).

2.4   Characterization of Drug Nanoparticles

There are various techniques used for characterization of drug nanoparticles. There is no single method that can be selected as the "best" for analysis. Most often the method is chosen to balance the restriction on sample size, information required, time constraints and the cost of analysis. Following methods are used commonly for characterization of drug nanoparticles. Different parameters and characterization methods for nanoparticles are shown in Table 2.1.

Table 2.1: Different parameters and characterization methods for nanoparticles

Parameters	Characterization method
Particle size and distribution	Photon correlation spectroscopy(PCS)
	Laser defractometry
	Transmission electron microscopy
	Scanning electron microscopy
	Atomic force microscopy
Surface hydrophobicity	Water contact angle measurement
	Rose Bengal(dye) binding
	X-ray photoelectron spectroscopy
Charge determination	Laser Doppler anemometry
	Zeta potentiometer
Carrier-drug interaction	Differential scanning calorimetry
Chemical analysis of surface	Static secondary ion mass spectrometry
	Sorptometer
Nanoparticle dispersion stability	Critical flocculation temperature(CFT)
Release profile	*In vitro* release characteristics under physiologic and sink conditions
Drug stability	Bioassay of drug extracted from nanoparticles
	Chemical analysis of drug

2.4.1   Particle size and size distribution

The characterization of particle size of nanosuspensions is done to obtain information about its average size, size distribution and change upon storage (e.g. crystal growth and/or agglomeration). Particle size distribution of drug nanoparticles can be measured using the following techniques:

2.4.2   Spectroscopy

As nanosuspensions usually comprises submicron particles, the appropriate method used to evaluate particle size distribution is photon correlation spectroscopy (PCS). In PCS or dynamic light scattering analyses scattered laser light from particles diffusing in a low viscosity dispersion medium (e.g. water). PCS analyses the fluctuation in velocity of the scattered light rather than the total intensity of the scattered light. The detected intensity signals (photons) are used to measure the correlation function. The diffusion coefficient D of the particles is obtained from the decay of this correlation function. The PI value is 0 in case if the particles are monodisperse. Incase of narrow distribution, the PI values vary between 0.10 and 0.20, values of 0.5 and higher indicate a very broad distribution (polydispersity). From the values of z-average and PI, even small increases in size of drug nanoparticles can be evaluated. The extent of increase in the particle size upon storage is a measure of instability. Therefore, PCS is considered as an sensitive instrument to detect instabilities during long-term storage (Kerker, 1969).

2.4.3   Laser diffraction

Laser diffractometry (LD) developed around 1980 is a very fast and used routinely in many laboratories. The instrument is also used for quantifying the amount of microparticles present, which is not possible using PCS. LD analyses the Fraunhofer diffraction patterns generated by particles in a laser beam. The first instruments were based on the Fraunhofer theory which is applicable for particle sizes 10 times larger than the wavelength of the light used for generating the diffraction pattern. For particle less than 6.3 μm (in case of using a helium neon laser, wavelength 632.8 nm) in size, the Mietheory is used to obtain the correct particle size distribution. Unfortunately, for most of pharmaceutical solids the refractive index is unknown. However, laser diffractometry is frequently used as a preferred characterization method for nanosuspensions because of its "simplicity" (Calvo, Vila-Jato and Alonso, 1996).

2.4.4   Microscopy

Microscopy based techniques can be used to study a wide range of materials with a broad distribution of particle sizes, ranging from nanometer to millimeter scale. Instruments used for microscopy based techniques include optical light microscopes, scanning electron microscopes (SEM) transmission electron microscopes (TEM) and atomic force microscopes (AFM). The choice of instrument for evaluation is determined by the size range of the particles being studied, magnification, and resolution. However, the cost of analysis is also observed to increase as the size of the particles decreases due to requirements for higher magnification, improved resolution, greater reliability and reproducibility. Optical microscopes tend to be more affordable and comparatively easier to operate and maintain than electron microscopes but have limited magnification and resolution (Molpeceres, Aberturas, and Guzman, 2000).

2.4.5   Differential scanning calorimetry (DSC)

Differential scanning calorimetry (DSC) is used to determine the crystallinity of drug nanoparticles by measuring its glass transition temperature, melting point and their associated enthalpies. This method along with X-ray powder diffraction (XRPD) described below is used to determine the extent to which multiple phases exist in the interior and their interaction following the milling process.

2.4.6   X-ray powder diffraction (XRPD)

X-ray powder diffraction (XRD) is a rapid analytical technique primarily used for phase identification of a crystalline material and can provide information on unit cell dimensions. X-ray diffraction is based on the constructive interference of monochromatic X-rays and a crystalline sample. These X-rays generated by a cathode ray tube are filtered to produce monochromatic radiation, collimated to concentrate, and directed toward the sample. The interference obtained is evaluated using Bragg's law to determine various characteristics of the crystal or polycrystalline material (Hunter, 1981).

2.5   Therapeutic Applications of Nanoparticles

2.5.1   Tumour targeting using nanoparticulate delivery systems

The rationale of using nanoparticles for tumour targeting is based on the nanoparticles will be able to deliver a concentrate dose of drug in the vicinity of the tumour targets via the enhanced permeability and retention effect or active targeting by ligands on the surface of nanoparticles; nanoparticles will reduce the drug exposure of health tissues by limiting drug distribution to target organ (Verdun et al., 1990).

2.5.2   Reversion of multidrug resistance in tumour cells

Multidrug resistance (MDR) is one of the most serious problems in chemotherapy. MDR occurs mainly due to the over expression of the plasma membrane pglycoprotein (Pgp), which is capable of extruding various positively charged xenobiotics, including some anticancer drugs, out of cells. Torestore the tumoural cells' sensitivity to anti cancer drugs by circumventing Pgp-mediated MDR, several strategies, including the use of colloidal carriers have been applied (Larsen, Escargueil and Skladanowski, 2000).

2.5.3   Nanoparticles for oral delivery of peptides and proteins

Polymeric nanoparticles allow encapsulation of bioactive molecules and protect them against enzymatic and hydrolytic degradation. The gastrointestinal tract provides a variety of physiological and morphological barriers against protein or peptide delivery, e.g. (a) proteolytic enzymes in the gut lumen-like pepsin, trypsin and chymotrypsin; (b) proteolytic enzymes at the brush border membrane (endopeptidases).

2.5.4   Nanoparticles for gene delivery

Several vaccines based nanomedicines functions to deliver genes to host cells and show their expression by production of antigenic protein to initiate immune response. One of the recent example of polynucleotide vaccines work by delivering genes encoding relevant antigens to host cells producing the antigenic protein within the vicinity of professional antigen presenting cells to initiate immune response. Such types of vaccines are responsible for the both humoral and cell-mediated immunity since intracellular production of protein, as opposed to extracellular deposition, stimulates both arms of the immune system (Panyam et al., 2002).

2.5.5   Nanoparticles for drug delivery into the brain

Nervous system is one of the most delicate microenvironments of the body which is protected by the blood–brain barrier (BBB) regulating its homeostasis. The blood–brain barrier (BBB) is the most important factor limiting the development of new drugs for the central nervous system. It has been discovered that the poly-(butylcyanoacrylate) based nanoparticles was able to deliver dalargin, hexapeptide, doxorubicin and other agents into the brain which is significant because of the great difficulty for drugs to cross the BBB.

2.5.6   Stem cell therapy

Nanotechnology presents efficient tools for improving stem cell therapy. The synergy between size, structure and physical properties of NPs makes them key players in revealing the fate and performance of stem cell therapy. Clearly NPs have much to offer in stem cell research and therapy. Stem cell therapies offer great potentials in the treatment for a wide range of diseases and conditions.

2.5.7   Gold nanoparticles detect cancer

Metallic nanostructures are more flexible particles compared to other nanomaterials owed to the possibility of controlling the size, shape, structure, composition, assembly, encapsulation and tunable optical properties. Between the metallic nanostructures possibly applied, AuNPs appears to be of great interest in the medical field, showing greater efficiency towards cancer therapy. Various researchers have utilized gold nanoparticles, such as ultrasensitive fluorescent probes to detect cancer biomarkers in human blood (Huang et al., 2007; Cobley et al., 2011).

2.6   Classification of Liposome

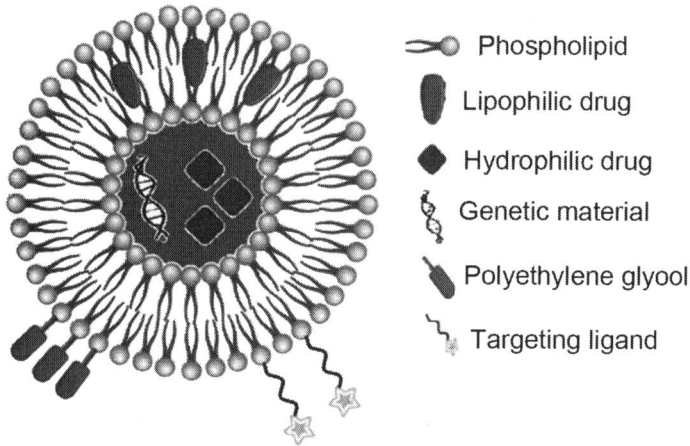

Figure 2.3: Schematic representation of liposome

Liposomes consist of enclosed vesicles of concentric self-assembling lipid bilayers comprises phospholipids and cholesterols in common (Nagle and Tristram-Nagle, 2000). Figure 2.3 represents basic structure of liposome with surface modification. According to the structure of lipid bilayers and the size of the vesicles, liposomes are commonly classified into small unilamellar vesicles (SUV), large unilamellar vesicles (LUV) and multilamellar vesicles (MLV) (Gregoriadis and Florence, 1993). Figure 2.4 represents the different types of liposome and there size range. The inner aqueous phase of liposomes is well protected by the lipid bilayers, and is able to load hydrophilic entities, whereas the hydrophobic region in the lipid bilayers is able to load hydrophobic entities (He et al., 2018). The liposome on the basis of delivery technology can be classified as conventional liposomes, pH sensitive liposomes, stealth liposomes, targeted liposome, stimuli-responsive liposome, gene-based liposome, immune liposomes, etc.

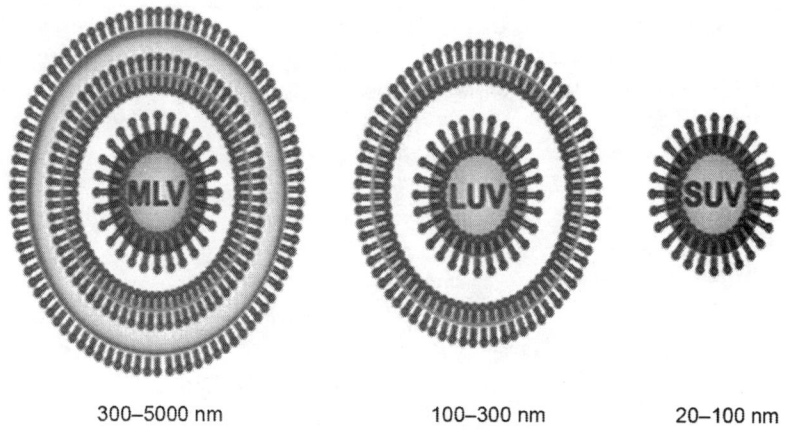

| 300–5000 nm | 100–300 nm | 20–100 nm |

Figure 2.4: Liposome types based on structure with structure modification

2.7 Materials Used in Liposomes

A wide range of phospholipids are available for preparation of liposomes. Phosphatidylcholines differ markedly from other phospholipids in the orientation of bilayer sheets with respect to the micellar structures (Lipowsky, 1991). A variety of natural and semi synthetic phospholipids are used in the preparation of liposomes. A comprehensive list of commonly used phospholipids is summarized in Table 2.2. The most commonly used natural phospholipids are egg and soya lecithin, obtained from egg and soya, respectively. Lipid composition and surface charge affect the tissue distribution and *in vivo* clearance of drug loaded liposomes. Neutral lipids consist of sphingomyelin or alkyl ether lecithin analogues in addition to phosphatidylcholine (Lasic, 1994). Increased resistance of lipids to hydrolysis without affecting the physical properties of the corresponding membrane can be achieved by replacing the ester with ether linkages. A variety of phospholipid structures can be obtained by combining the polar head groups (e.g. phosphatidylcholilne (PC), phsophatidylethanolamine (PE), phosphatidylserine (PS), phosphatidylglycerol (PG) and phosphatidic acid (PA)) with various fatty acids chains, such as lauric, myristic, palmitic, stearic and oleic acids. Similarly, cardiolipin (CL) and sphingomyelin (SM) combined with these fatty acid chains provide additional lipids for specialized drug delivery (Vemuri and Rhodes, 1995).

Table 2.2: Classification of component of liposome with example

Natural phospholipids	Synthetic phospholipids	Sphingolipids
Phosphatidyl choline (Lecithin) Phosphatidyl ethanolamine (cephalin) Phosphatidyl serine Phosphatidyl inositol Phosphatidyl Glycerol	Dipalmitoyl phosphatidyl choline Distearoyl phosphatidyl choline Dipalmitoyl phosphatidyl ethanolamine Dipalmitoyl phosphatidyl serine Dipalmitoyl phosphatidic acid Dipalmitoyl phosphatidyl glycerol Dioleoyl phosphatidyl choline Dioleoyl phosphatidyl glycerol	Sphingomyelin Glycosphingo lipids Gangliosides— found on grey matter

2.8   Liposome Preparation Methods

The following methods are used for the preparation of liposome:

1. Mechanical dispersion method.
2. Solvent dispersion method.
3. Detergent removal method (removal of nonencapsulated material

2.8.1   Mechanical dispersion method

2.8.1.1   Sonication

Sonication is perhaps the most extensively used method for the preparation of SUV. Here, MLVs are sonicated either with a bath type sonicator or a probe sonicator under a passive atmosphere. The main disadvantages of this method are very low internal volume/encapsulation efficacy, possible degradation of phospholipids and compounds to be encapsulated, elimination of large molecules, metal pollution from probe tip, and presence of MLV along with SUV. In case of probe sonication, the tip of a sonicator is directly engrossed into the liposome dispersion. In case of bath sonication, the liposome dispersion in a cylinder is placed into a bath sonicator (Maurer, Fenske and Cullis, 2001).

2.8.1.2   French pressure cell extrusion

French pressure cell involves the extrusion of MLV through a small orifice. An interesting comment is that French press vesicle appears to recall entrapped solutes significantly longer than SUVs do, produced by sonication or detergent removal. The drawbacks of the method are that the high temperature is difficult to attain, and the working volumes are comparatively small (about 50 mL as the maximum) (Nekkanti and Kalepu, 2015).

2.8.1.3   Freeze–thawed liposomes

SUVs are rapidly frozen and thawed slowly. The short-lived sonication disperses aggregated materials to LUV. The creation of unilamellar vesicles is as a result of the fusion of SUV throughout the processes of freezing and thawing. The encapsulation efficacies from 20% to 30% were obtained (Llu and Yonetani, 1994).

2.8.2   Solvent dispersion method

2.8.2.1   Ether injection (solvent vaporization)

A solution of lipids dissolved in diethyl ether or ether–methanol mixture is gradually injected to an aqueous solution of the material to be encapsulated at 55°C to 65°C or under reduced pressure. The consequent removal of ether under vacuum leads to the creation of liposomes. The main disadvantages of the technique are that the population is heterogeneous (70 to 200 nm) and the exposure of compounds to be encapsulated to organic solvents at high temperature (Allen and Cullis, 2013; Zylberberg and Matosevic, 2016).

2.8.2.2   Ethanol injection method

A lipid solution of ethanol is rapidly injected to a huge excess of buffer. The MLVs are at once formed. The disadvantages of the method are that the population is heterogeneous (30 to 110 nm), liposomes are very dilute, the removal all ethanol is difficult because it forms into a zoetrope with water (Nekkanti and Kalepu, 2015).

2.8.2.3   Reverse-phase evaporation method

This method provided a progress in liposome technology, since it allowed for the first time the preparation of liposomes with a high aqueous space-to-lipid ratio and a capability to entrap a large percentage of the aqueous material presented. Reverse-phase evaporation is based on the creation of inverted micelles. The slow elimination of the organic solvent leads to the conversion of these inverted micelles into viscous state and gel form (Mufamadi et al., 2011; He et al., 2018).

2.8.3   Detergent removal method (removal of nonencapsulated material)

2.8.3.1   Dialysis method

The detergents at their critical micelle concentrations (CMC) have been used to solubilize lipids. As the detergent is detached, the micelles become increasingly better-off in phospholipid and lastly combine to form LUVs. The detergents were removed by dialysis (Schmidtgen et al., 1998).

2.8.3.2   Detergent absorption method

Detergent absorption is attained by shaking mixed micelle solution with beaded organic polystyrene adsorbers. The great benefit of using detergent adsorbers is that they can eliminate detergents with a very low CMC, which are not entirely depleted (Schmidtgen et al., 1998).

2.8.3.3   Gel-permeation chromatography method

In this method, the detergent is depleted by size special chromatography. The liposomes do not penetrate into the pores of the beads packed in a column. The pretreatment is done by pre saturation of the gel filtration column by lipids using empty liposome suspensions (Jederström and Russell, 1981; Andrieux et al., 1998).

2.8.3.4   Dilution method

Upon dilution of aqueous mixed micellar solution of detergent and phospholipids with buffer and as the system is diluted beyond the mixed micellar phase boundary, a spontaneous transition from polydispersed micelles to vesicles occurs (Fry, White and Goldman, 1978).

2.9   Characterization of Liposomes

Liposomes produced by different methods have varying physicochemical characteristics, which leads to differences in their *in vitro* (sterilization and shelf life) and *in vivo* (disposition) performances. Rapid, precise and reproducible quality control tests are required for characterizing the liposomes after their formulation and upon storage for a predictable *in vitro* and *in vivo* behaviour of the liposomal drug product. A liposomal drug product can be characterized for some of the parameters that are discussed below.

2.9.1   Size and size distribution

The size distribution is of primary consideration, since it influences the *in vivo* fate of liposomes along with the encapsulated drug molecules. Various techniques to determine the size of the vesicles, include microscopy (optical microscopy, negative stain transmission electron microscopy, cryo-transmission electron microscopy, freeze fracture electron microscopy and scanning electron microscopy), diffraction and scattering techniques (laser light scattering and photon correlation spectroscopy) and hydrodynamic techniques (field flow fractionation , gel permeation and ultracentrifugation) (Katare, Vyas and Dixit, 1991; Moon and Giddings, 1993; Schmidtgen et al., 1998).

2.9.2   Percentage drug encapsulation

The amount of drug encapsulated/entrapped in liposome vesicle is given by percentage drug encapsulation. Column chromatography can be used to estimate the percentage drug encapsulation of liposomes. Then, the fraction of liposomes containing the encapsulated drug is treated with a detergent, so as to attain lysis, which leads to the discharge of the drug from the vesicles into the surrounding medium. This exposed drug is assayed by a suitable technique which gives the percent drug encapsulated from which encapsulation efficiency can be calculated (Fry, White and Goldman, 1978; Jederström and Russell, 1981; Betageri, 1993; Andrieux et al., 1998).

2.9.3   Surface charge

Since the charge on the liposome surface plays a key role in the *in vivo* disposition and stability, it is essential to know the surface charge on the vesicle surface. Two methods; namely, free-flow electrophoresis and zeta potential measurement can be used to estimate the surface charge of the vesicle (Dan, 2002).

2.9.4   Vesicle shape and lamellarity

Various electron microscopic techniques can be used to assess the shape of the vesicles. The number of bilayers present in the liposome, i.e. lamellarity can be determined using freeze fracture electron microscopy and nuclear magnetic resonance analysis. Apart from knowing the shape and lamellarity, the surface morphology of liposomes can be assessed using freeze–fracture and freeze–etch electron microscopy (Llu and Yonetani, 1994).

2.9.5   Phospholipid identification and assay

The chemical components of liposomes must be analyzed prior to and after the preparation. Barlett assay, Stewart assay and thin layer chromatography can be used to estimate the phospholipid concentration in the liposomal formulation.

2.9.6   Stability study of liposomes

During the development of liposomal drug products, the stability of the developed formulation is of major consideration. Hence a stability protocol is essential to study the physical and chemical integrity of the drug product in its storage.

2.10   Therapeutic Applications of Liposome

2.10.1   Site-avoidance delivery

The cytotoxicity of anti-cancer drugs to normal tissues is attributed to their narrow therapeutic index (TI). Under such circumstances, the TI can be improved by minimizing the delivery of drug to normal cells by encapsulating in liposomes (Alyane, Barratt and Lahouel, 2016).

2.10.2   Site-specific targeting

Delivery of a larger fraction of the drug to the desired (diseased) site, reducing the drug's exposure to normal tissues can be achieved by site-specific targeting. On systemic administration, long circulating immunoliposomes are bind to target cells with greater specificity (Paszko and Senge, 2012).

2.10.3   Intracellular drug delivery

Increased delivery of poorly absorbed drugs to the cytosol can be accomplished using liposome. The drugs are encapsulated within liposomes, showed greater activity in comparison to free drug (Krieger et al., 2010).

2.10.4   Sustained release drug delivery

To achieve the optimum therapeutic efficacy, which requires a prolonged plasma concentration at therapeutic level; liposomes provide sustained release of target drugs. Drugs can be encapsulated in liposomes for sustained release and optimized drug release rate can be optimized (Loira-Pastoriza, Todoroff and Vanbever, 2014).

2.10.5   Immunological adjuvants in vaccines

Liposomes can be used for enhancing the immune response by encapsulating the adjuvants. Depending on the lipophilicity of antigens, the liposome can accommodate antigens in the aqueous cavity or incorporate within the bilayers. These targeting ligands could be, vitamins, specific antigens or monoclonal antibodies (Chaudhary, Kohli and Kumar, 2013).

2.11   Conclusions

To overcome the limitations of conventional drug delivery system, continuous steady rate of drug at the site of action targeted drug delivery systems was developed it requires very small amount of drug comparatively to conventional dosage forms. For decades pharmaceutical sciences have been using nanoparticles as targeted drug delivery to reduce toxicity and side effects of drugs. A conceptual understanding of structure ad types of nanomaterials is needed to develop and apply safely in drug delivery in the future. Furthermore, characterization of nanoparticle is an important aspect of predicting its physiological behaviour after administration. Owing to this, the toxicity inside the body and side effects of the drug on organs are reduced. Drugs can be delivered to the exact location with the right amount of dosage. Liposomes have been in the use as drug delivery systems in the recent years with a few formulations commercially available. Liposomes are classified according to structure and delivery technology. This can be prepared by different production method as explained above. Liposomes as a drug delivery system include benefits like improved pharmacokinetics and pharmacodynamics, decreased toxicity, enhanced therapeutic efficacy against pathogens and improved drug-target selectivity.

2.12   References

Allen, T.M., and Cullis, P.R. (2013), 'Liposomal drug delivery systems: From concept to clinical applications', *Adv. Drug Del. Rev.*, 65(1), 36–48.

Alyane, M., Barratt, G. and Lahouel, M. (2016), 'Remote loading of doxorubicin into liposomes by transmembrane pH gradient to reduce toxicity toward H9c2 cells', *Saudi Pharm. J.*, 24(2), 165–175.

Andrieux, K., et al. (1998), 'Methodology for vesicle permeability study by high-performance gel exclusion chromatography', *J. Chromatograph. B. Biomed. Sci. Appl.*, 706(1), 141–147.

Balazs, D.A., and Godbey, W. (2011), 'Liposomes for use in gene delivery', *J. Drug Del.*, 2011, 1–12.

Bangham, A.D., and Horne, R.W. (1964), 'Negative stainingof phospholipids and their structural modificationby surface-active agents as observed in the electron microscope', *J. Mol. Biol.*, 8, 660–668.

Bardania, H., Tarvirdipour, S. and Dorkoosh, F. (2017), 'Liposome-targeted delivery for highly potent drugs', *Artif. Cells, Nanomed. Biotechnol.*, 45(8),1478–1489.

Barichello, J.M., et al. (1999), 'Encapsulation of hydrophilic and lipophilic drugs in PLGA nanoparticles by the nanoprecipitation method. *Drug Dev. Ind. Pharm.*, 25(4), 471–476.

Betageri, G.V. (1993), 'Liposomal encapsulation and stability of dideoxyinosine triphosphate', *Drug Devel. Industr. Pharm.* 19(5), 531–539.

Calvo, P., Vila-Jato, J.L. and Alonso, M.J. (1996) 'Comparative in vitro evaluation of several colloidal systems, nanoparticles, nanocapsules, and nanoemulsions, as ocular drug carriers', *J Pharmaceut. Sci.*, 85(5), 530–536.

Chaudhary, H., Kohli, K. and Kumar, V. (2013), 'Nano-transfersomes as a novel carrier for transdermal delivery', *Int. J. Pharmaceut.*, 454(1), 367–380.

Cobley, C.M., et al. (2011), 'Gold nanostructures: A class of multifunctional materials for biomedical applications', *Chem. Soc. Rev.*, 40(1), 44–56.

Dan, N. (2002), 'Effect of liposome charge and PEG polymer layer thickness on cell-liposome electrostatic interactions', *Biochim Biophys. Acta*, 1564(2), 343–348.

Fry, D.W., White, J.C. and Goldman, I.D. (1978), 'Rapid separation of low molecular weight solutes from liposomes without dilution', *Anal. Biochemist.*, 90(2), 809–815.

Gregoriadis, G., and Florence, A.T. (1993), 'Liposomes in drug delivery', *Drugs*, 45(1), 15–28.

Haisheng, H., Yi, Lu et al. (2019), 'Adapting liposomes for oral drug delivery', *Acta Pharm. Sin B.*,9(1), 36-48.

Huang, X., Jain, P.K., El-Sayed, I.H., El-Sayed, M.A. (2007), 'Gold nanoparticles: Interesting optical properties and recent applications in cancer diagnostics and therapy', *Nanomedicine*, 2(5), 681–693.

Hunter, R. J. (1988), *Zeta potential in colloid science : Principles and applications*, Academic Press, London.

Molpeceres, J., Aberturas, M.R., and Guzman, M. (2000), 'Biodegradable nanoparticles as a delivery system for cyclosporine: Preparation and characterization', *J. Microencapsul.*, 17(5), 599–614.

Jederström, G., and Russell, G. (1981), 'Size exclusion chromatography of liposomes on different gel media', *J. Pharm. Sci.*, 70(8), 874–878.

Jeon, H.J., et al. (2000), 'Effect of solvent on the preparation of surfactant-free poly(DL-lactide-*co*-glycolide) nanoparticles and norfloxacin release characteristics', *Int. J. Pharm.*, 207(1–2), 99–108.

Katare, O.P., Vyas, S.P. and Dixit, V.K. (1991), 'Proliposomes of indomethacin for oral administration', *J. Microencapsul.*, 8(1), 1–7.

Kerker, M. (1969), *The scattering of light : And other electromagnetic radiation,* Academic Press, London.

Klose, D., Siepmann, F., Willart, J.F., Descamps, M., Siepmann, J. (2010), 'Drug release from PLGA-based microparticles: Effects of the "microparticle: Bulk fluid" ratio', *Int. J. Pharm.*, 383(1–2), 123–131.

Krieger, M.L., Eckstein, N., Schneider, V., Koch, M., Royer, H.D., Jaehde, U., Bendas, G. (2010), 'Overcoming cisplatin resistance of ovarian cancer cells by targeted liposomes in vitro', *Int. J. Pharm.*, 389(1–2), 10–17.

Larsen, A.K., Escargueil, A.E. and Skladanowski, A. (2000), 'Resistance mechanisms associated with altered intracellular distribution of anticancer agents.', *Pharmacol Therap.*, 85(3), 217–229.

Lasic, D.D. (1993), *Liposomes: From Physics to Applications,* Elsevier, New York, Amsterdam.

Lipowsky, R. (1991), 'The conformation of membranes', *Nature*, 349(6309), 475–481.

Llu, L., and Yonetani, T. (1994) 'Preparation and characterization of liposome-encapsulated haemoglobin by a freeze–Thaw method', *J. Microencapsul.*, 11(4), 409–421.

Loira-Pastoriza, C., Todoroff, J. and Vanbever, R. (2014), 'Delivery strategies for sustained drug release in the lungs', *Adv. Drug Del. Rev.*, 75, 81–91.

Maurer, N., Fenske, D.B. and Cullis, P.R. (2001), 'Developments in liposomal drug delivery systems', *Exp. Opi. Biol. Ther..*, 1(6), 923–947.

Moon, M.H., and Giddings, J.C. (1993), 'Size distribution of liposomes by flow field-flow fractionation', *J. Pharm. Biomed. Anal.*, 11(10), 911–920.

Mufamadi, M.S., et al. (2011), 'A review on composite liposomal technologies for specialized drug delivery', *J. Drug Del.*, 2011, 1–19.

Nagle, J.F., and Tristram-Nagle, S. (2000), 'Structure of lipid bilayers', *Biochim. et Biophys. Acta*, 1469(3), 159–195.

Nekkanti, V., and Kalepu, S. (2015), 'Recent advances in liposomal drug delivery: A review', *Pharm. Nanotechnol.*, 3(1), 35–55.

Niwa, T., et al. (1993), 'Preparations of biodegradable nanospheres of water-soluble and insoluble drugs with d,l-lactide/glycolide copolymer by a novel spontaneous emulsification solvent diffusion method, and the drug release behavior', *J. Controll. Rel.*, 25(1–2), 89–98.

Panyam, J., et al. (2002), 'Rapid endo-lysosomal escape of poly(dl -lactide- co -glycolide) nanoparticles: Implications for drug and gene delivery', *FASEB J.*, 16(10), 1217–1226.

Paszko, E., and Senge, M.O. (2012), 'Immunoliposomes', *Curr. Med. Chem.*, 19(31), 5239–5277.

Pinto Reis, C., et al. (2006), 'Nanoencapsulation I. Methods for preparation of drug-loaded polymeric nanoparticles', *Nanomed. Nanotechnol. Biol. Med.*, 2(1), 8–21.

Puri, A., et al. (2009), 'Lipid-based nanoparticles as pharmaceutical drug carriers: From concepts to clinic', *Crit. Rev. Therap. Drug Carr. Syst.*, 26(6), 523–580.

Quintanar-Guerrero, D., et al. (1998), 'Preparation techniques and mechanisms of formation of biodegradable nanoparticles from preformed polymers', *Drug Dev. Industr. Pharm.*, 24(12), 1113–1128.

Schmidtgen, M. C., Drechsler, M., Lasch, J., Schubert, R. (1998), 'Energy-filtered cryotransmission electron microscopy of liposomes prepared from human stratum corneum lipids', *J. Micro.*, 191(2), 177–186.

Soppimath, K.S., et al. (2001), 'Biodegradable polymeric nanoparticles as drug delivery devices', *J. Controll. Rel.*, 70(1–2), 1–20.

Vemuri, S., and Rhodes, C. (1995), 'Preparation and characterization of liposomes as therapeutic delivery systems: A review', *Pharmaceut. Acta Helvet.*, 70(2), 95–111.

Verdun, C., et al. (1990), 'Tissue distribution of doxorubicin associated with polyisohexylcyanoacrylate nanoparticles', *Cancer Chemother. Pharmacol.*, 26(1), 13–18.

York. (1999), 'Strategies for particle design using supercritical fluid technologies', *Pharmaceut. Sci. Technol Today*, 2(11), 430–440.

Zylberberg, C., and Matosevic, S. (2016), 'Pharmaceutical liposomal drug delivery: A review of new delivery systems and a look at the regulatory landscape', *Drug Del.*, 23(9), 3319–3329.

3

Microcapsules/Microspheres: Types, Preparation and Evaluation; Monoclonal Antibodies; Preparation and Application, Niosomes, Aquasomes, Phytosomes, Electrosomes; Preparation and Application

Rabinarayan Parhi[*1], Suryakanta Swain[2], Sambamoorthy Unnam[3] and Sitty Manohar Babu[2]

[1] *Department of Pharmaceutics, Gitam Institute of Pharmacy, GITAM University (Deemed to be University), Gandhinagar campus, Rushikonda, Visakhapatnam-530045, Andhra Pradesh, India*

[2] *Southern Institute of Medical Sciences, College of Pharmacy, Department of Pharmaceutics, Guntur-522001, Andhra Pradesh, India*

[3] *Department of Pharmaceutics, NRI College of Pharmacy, NRI Group of Institutions, Pothavarappadu (V), Via Nunna, Agiripalli (M), Vijayawada Rural, Krishna District, Andhra Pradesh-522212, India*

3.1   Introduction

To improve and simplify lifestyle of human being, and to reduce economic burden on patient are the main yardstick which necessitates the use of advance technology in pharmaceutical field. In this context, drug delivery system (DDS) came a long way to fulfil the expectation of population at large. Delivering the biological actives in controlled manner for considerable period of time and to maintain drug level in the body within therapeutic window (provide safety) are the main objectives with which DDS was developed. Controlled drug delivery systems (CDDS) have been developed to not only address the above aspects, but also to deliver the drug to the target site (targeted drug delivery). Micro- and nanocarriers, in their various forms, with the advancement in technology, have the potential to provide endless opportunities in the area of drug delivery. Particulate systems offers many advantages: (i) enhance bioavailability by improving their bioavailability, (ii) increase residence time in the body by decreasing its clearance, (iii) target specific site of the body, (iv) deliver less side effects, and (v) deliver drugs with poor water solubility (Coelho JF at al., 2010; Mudshinge SR et al., 2011).

Among all the particulate systems, microspheres are one of the most common types and have many advantages. Microspheres can encapsulate various types of drugs, including vaccines, antibiotics, protein and nucleic acids. In addition, microcapsules can be administered through various routes, such as oral, topical and parenteral. They are generally biodegradable and biocompatible, and can offer high bioavailability and have potential to sustain the drug release for desired period of time. They have the ability to improve patient compliance by replacing the frequent dose with infrequent (once in a month or less) commercial injectable microspheres, such as Nutropin Depot® and LupronDepot® (Kim K K and Pack D W, 2006).

With the major advancement in genetic sequencing and biomedical research, the medicine progresses into a new era of personalized therapy. It will create more unified treatment approach specific to the genome and the individual. In this context, much research into monoclonal antibodies (mAbs) focuses on identifying new targets in order to develop and improve their efficacy for the use in clinical setting. The first mAb was approved licence in 1986 was chimeric one. Techniques, such as genetic sequencing made it possible to move from chimeric having side effects to human mAbs without any side effects. Presently, there are approximately 30 mAbs which have been approved by FDA for their use in clinical setting despite of so many hurdles including, lack of efficacy, issue of affordability and the use of inefficient models for generation. These mAbs are used to treat various conditions and diseases, such as cancer, transplant rejection, chronic anti-inflammatory diseases, cardiovascular diseases, respiratory tract related disease and infectious disease (Liu JKH, 2014).

When compared with all, vesicular DDS with a bilayer membrane and a hollow space inside have been at the center of great attention because of their ability to encapsulate both hydrophilic and hydrophobic drugs, high storage time, nontoxic, can be used in various routes, receptive surface for treating various targeting agents. Niosome is one class of vesicular nanocarriers which is composed of nonionic surfactant, cholesterol or other amphiphilic molecules. Owing to the presence of non-ionic surfactant, niosomes show more stability when compared with liposomes. In addition, niosomes require less production cost, highly pure and offer suitable surface chemistry that can modify the encapsulated drug's intrinsic pharmacokinetics and eventual drug targeting (Moghassemi S and Hadjizadeh A, 2014; Marianecci C et al., 2014).

Aquasomes are one of the most recently developed selfassembled nanoparticulate drug carries. Unlike simple nanoparticles, they are three layered structures with a core of solid phase nanocrystal covered with a layer of oligomeric film to which biochemically actives are adsorbed with or without modification. These are developed with initial intention of delivering peptides, proteins, insulin, enzymes, antigen and genes to desired site in the body. The absence of interaction between drug and carrier, and extended drug stability provided by the oligosaccharide coating are the advantages of aquasomes (Jain SS et al., 2012; Banerjee S and Sen KK, 2018).

Phytosomes are another vesicular carrier which is developed by reacting phospholipids with selected phytoconstituents with a suitable solvent. These are considered as advanced form of herbal formulations that have higher capability to carry the medicinal plant of hydrophilic agent through the lipid bilayer and therefore, show higher bioavailability and antioxidant effect of encapsulated phytoconstituents when compared with conventional dosage forms as well as liposomes (Karimi N et al., 2015). The present chapter discusses up to date information on types, methods of preparation, evaluation and applications of particulate systems, such as microspheres, niosomes, aquasomes, and phytosomes along with mAbs. In addition, the concept of electrosomes is also mentioned.

3.2   Micro Capsules/Micro Spheres

Microparticles are defined as solid particles that have size range from 1 to 1000 μm and made of polymeric, proteinic and waxy materials (Karmakar U and Faysal MM, 2009). Microparticles are generally classified in to two types: microcapsules and microspheres. In microcapsules, the macromolecular materials with the active principles are completely surrounded by a distinct capsular wall or membrane (Figure 3.1). Whereas, microspheres are matrix systems with the active ingredients dispersed throughout (Deasy PB, 1984; Edith M and Mark RK, 1998). Therefore, microspheres are having active ingredients embedded in a polymeric or protein matrix network in either molecular dispersion or a solid aggregated state and in microcapsules the active ingredients are coated by a solidified polymeric or proteinic envelope (Ganesan P et al., 2014).

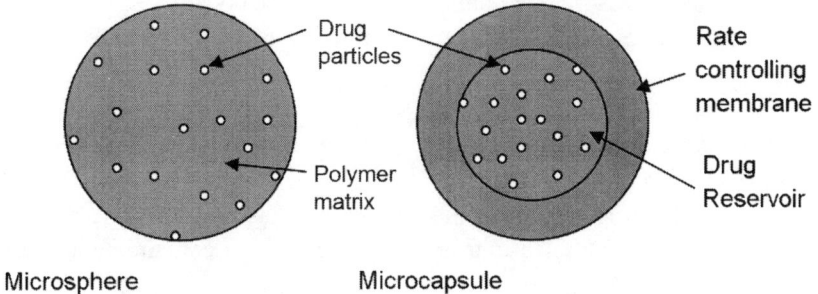

Microsphere   Microcapsule

Figure 3.1: Schematic illustration of microsphere and microcapsule

Ideal properties of microspheres: Microspheres should satisfy following criteria to become ideal (Sahil K et al., 2011):

- Control the drug release without modifying its normal bio-fate in the body after administration.

- Improve therapeutic efficacy.
- Incorporate reasonably high concentrations of active ingredients.
- Stability of the preparation with acceptable self-life.
- Biocompatibility along with desired biodegradability.
- Minimum or no toxicity.
- Controlled particle size and easy dispersion for the parenteral administration.
- Sterilizability.

Microspheres as DDS is very popular as they possesses several advantages (Deasy PB, 1984; Edith M and Mark RK, 1998; Meena K P et al., 2011; Urs AVR et al., 2010)

- Reliable means of site specific drug delivery, i.e. targeted drug delivery with reduced side effects.
- Sustained release of active ingredients from biodegradable microspheres.
- Reduce the dose size thereby the cost of the product and therapy.
- Reduces dosing frequency therefore, promote patient compliance.
- Reliable means of site specific drug delivery, i.e. targeted drug delivery with reduced side effects.
- Provide protection to unstable drugs both before and after administration.
- Allow manipulation in tissue distribution, cellular interaction of the drug, pharmacokinetic profile and *in vivo* action of drug.
- Reduced gastric irritation and first-pass metabolism.
- Improve biological half-life therefore, the bioavailability.
- Can be used in the parenteral routes due to small size and spherical shape.

There are certain disadvantages of microspheres including (Thanou M et al., 2001; Bansal H et al., 2011)

- Any loss of integrity of the dosage form may lead to dose dumping as controlled release formulations generally contain higher drug loading.
- There are possibilities of variety of factors including intrinsic and extrinsic influences the drug release in controlled manner.
- Variability in drug release from one dose to another dose.
- Microspheres, when administered parenterally may interact or form complexes with blood component.
- Failure of parenteral microspheres may lead to severe toxic or adverse effect which cannot be reversed.

3.2.1   Types of Microspheres

There are five types of microspheres described in the literature, including bioadhesive, mucoadhesive, floating, magnetic and radioactive microspheres (Figure 3.2).

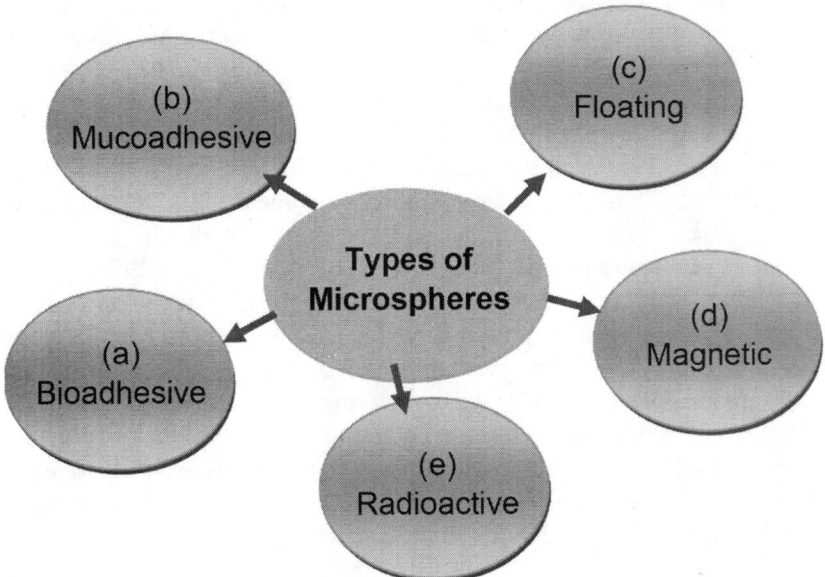

Figure 3.2: Schematic representation of different types of microspheres

3.2.1.1   Bioadhesive microspheres

Adhesion is defined as the state in which two surfaces are held together by interfacial forces involving either interlocking action or valence forces or both (Kinloch A, 1982). Bioadhesion is a specific case of adhesion wherein two substrates, at least one biological in nature, such as buccal, nasal, ocular, rectal, skin etc., are held together for an extended period of time due to above forces (Jimenez-Castellanos MR et al., 1993). In the context of DDS, bioadhesion has the meaning of sticking of drug to the biological membrane by virtue of adhesion property of water-soluble polymer. Bioadhesive microsphere DDS allows an intimate and prolonged contact at the target site to form between incorporated drugs and biological tissues. This contact may lead to higher drug absorption and bioavailability, localized drug delivery in the desired regions, such as skin and various body cavities, modification of permeability of mucosal tissue or membranes resulted in enhanced adsorption of micro- and macro-molecules, and providing sustained or controlled release of therapeutic agents, thereby reducing frequency of application and patient compliance (Khutoryanskiy VV, 2011; Senthil A et al., 2011; Patel JK et al., 2010).

3.2.1.2   Mucoadhesive microspheres

Mucoadhesion is defined as the attachment of drug carrier to the mucosal layer. It involves three general steps, such as wetting, adsorption and interpenetration of polymer chains with that of mucin. Mucoadhesion is a complex process and many theories have been proposed to understand the mechanism of adhesion including (i) the wetting theory (Gu JM et al., 1998; Shaikh R et al., 2011), (ii) the electronic theory (Carvalho FC et al., 2010; Dodou D et al., 2005), (iii) adsorption theory (Jimenez-Castellanos MR et al., 1993; Ahagon A and Gent A, 1975), (iv) the diffusion theory (Vasir JK et al., 2003), (v) fracture theory (Jimenez-Castellanos MR et al., 1993; Gu JM et al., 1998), and (vi) the mechanical interlocking theory (Carvalho FC et al., 2010).Mucoadhesive microspheres of diameter ranging from 1 to 1000 μm composed either entirely of mucoadhesive polymer or having an outer coating of it. Appropriately tailored mucoadhesive microspheres can exhibit many advantages including specific targeting of drug to the absorption site, efficient absorption leading to higher bioavailability, localized as well as systemic absorption of drug in controlled manner. They can be applied to any mucosal tissue those found in various body cavities and eye (Shadab Md et al., 2012).

3.2.1.3   Floating microspheres

The density of gastric content is 1.004 g/cm^3. Therefore, any DDS having bulk density less than this value could be used as floating systems. Floating microspheres have bulk density < 1 g/cm^3 and remain buoyant in the stomach for a prolonged period of time and thereby reduces dosing frequency (Bardonnet PL et al., 2006). These microspheres avoid the variation in gastric emptying rates and release the drug at desired rate with reduced chances of dose dumping. Floating microspheres are of two types: effervescent and noneffervescent. Effervescent microspheres are matrix types of systems which liberate CO_2 gas and get entrapped in swollen polymer leading to lowering of density of the total system. This provides buoyancy to the microspheres for a prolonged period of time along with the controlled drug release. These systems composed of swellable polymers (i.e. chitosan and methylcellulose) and effervescent materials, such as sodium biocarbonate, tartaric acid and citric acid (Nasa P et al., 2010). On the other hand noneffervescent microspheres are prepared with gel forming or swellabe cellulose type of hydrocolloids, polysaccharides and matrix forming polymers (i.e. polyacrylate, polystyrene). Upon contact with gastric fluid these microspheres swell up and attained a bulk density less than 1 g/cm^3. This resulted in entrapment of air within these microspheres and imparts buoyancy to it. The so-formed swollen structure acts as a reservoir and releases the drug in sustained manner (Arora S et al., 2005).

3.2.1.4 Magnetic microspheres

An ideal DDS delivers the drug in controlled manner and then localizes the drug to the target site. In this context, magnetic microspheres composed of polymer (e.g., chitosan, dextran), magnetic material (such as supra-magnetic iron oxide) and drug or therapeutic isotopes have the ability to localize the drug at disease site (Solanki N, 2018). This not only reduce the distribution of drug to all body parts which is the root cause for side effects, but also reduces dose of the drug by several fold. Upon parenteral administration of magnetic microspheres, a powerful magnetic field is placed just above the affected area to localize the magnetic microspheres and thereby releasing the drug in higher amount (Maestrelli F et al., 2008).These microspheres can be used in the treatment of cancer where site specific drug delivery is a prerequisite to minimize side effects. Magnetic microspheres can be classified in to two types: therapeutic and diagnostic microspheres. The former used deliver chemotherapeutic agents or therapeutic radioisotopes to different target site. In case of cancer, it can avoid the use of external beam therapy where more chances of cell damage is imminent. Diagnostic microspheres act as contrast agents for magnetic resonance imaging. Smaller supra-magnetic iron oxides of nanometer sizes are generally used for the diagnosis of liver metastasis and other abdominal structures (Najmuddin M et al., 2010).

3.2.1.5 Radioactive microspheres

Microspheres containing radioisotopes are called as radioactive microspheres. It is different from DDS as the radioactivity is not released from the microspheres instead from the encapsulated radioisotopes. Based on the incorporated isotopes, the different types of radioactive microspheres are α, β, and γ emitters (Yadav AV and Mote HH, 2007). These are used in patients suffering from cancers, diabetic ulcers and other diseases. In case of cancer treatment, the microspheres larger than the diameter of capillaries (10-30nm) get trapped in first capillary bed of arteries after injected into arteries and lead to tumour treatment of interest (Krishna KVM et al., 2013). Apart from treatment these microspheres also used in diagnosis including imaging of tumour, thrombus in deep vein, bone marrow, liver and spleen, gated blood pool study, blood flow measurement, lung scintigraphy and investigation of biodistribution of microspheres (Heymann MA et al., 1997).

3.2.2   Preparation of Microcapsule/Microspheres

Presently, there are many methods available to prepare microspheres. However, the choice of technique is mainly depends on following criteria including, (i) the physical, chemical and biological activity of encapsulated drugs to be maintained, (ii) microspheres should possesses reasonable size range particularly for parenteral administration, (iii) method should yield microspheres with high encapsulation efficiency, (iv) there should not be any toxic remnant in the finished product, (v) microspheres should release the drug in predetermined rate without burst release, and (vi) the selected method should produce the microspheres in large scale (Vyas SP and Khar RK, 2002). Following methods are more commonly used to prepare microspheres;

3.2.2.1   *Coacervation technique*

This technique involves the dispersion of an aqueous drug solution in a solution of water immiscible solvent and polymer. Subsequently, the polymer gets deposited on the surface of the aqueous drug droplets in the form of layer upon evaporation of water immiscible volatile solvent. Thus, the technique depends on the principle of reducing the solubility of the polymer in the organic phase to induce the formation of polymer rich phase termed as coacervates. The hydrophilic drugs are dissolved in aqueous medium, whereas lipophilic drugs are dissolved in the polymeric solution. Then, by changing the solution condition phase separation is accomplished. Using coacervation method, water soluble drugs, such as protein, peptides and lipophilic drugs, such as steroids can easily be encapsulated in microspheres. Coacervation can also be achieved by thermal change method. This is carried out by dissolving water insoluble polymer, such as ethyl cellulose in suitable solvent (e.g. cyclohexane) with vigorous stirring at 80°C by heating. Then, the grinded drugs are added to above solution with vigorous stirring and phase separation was induced by reducing temperature or using ice bath. Afterwards, the product was washed with the above solvent and air dried followed by passing them through sieve No. 40. (Singh A et al., 2012).

3.2.2.2   *Hot melt microencapsulation*

As the name indicates, this method involves mixing of already sieved drug molecules with the melted polymer. Subsequently, the above mixture is suspended in a nonmiscible solvent, such as silicone oil at temperature 5°C above the melting point of the polymer with continuous stirring (Ramesh DV, 2009). After stabilization of the mixture, the microspheres are solidified by cooling. This is followed by washing of microspheres with suitable solvent, such as petroleum ether (Singh A et al., 2012).

3.2.2.3   Spray drying and spray congealing

Main concept of both the methods is drying of the fine droplets composed of drug and polymer in the air stream. Based on the drying principle, i.e. removal of solvent by heating or cooling of the droplets, two methods are named as spray drying and spray congealing, respectively. Therefore, the principle difference between the two methods is the means by which solidification of droplets is accomplished. The polymer is dissolved in the suitable organic solvents, such as acetone or dichloromethane and then the drug is dispersed in the polymeric solution under high speed homogenization. Th   e resulted dispersion is then atomized in a spray chamber under a stream of hot air which supplies the latent heat of vaporization required to remove the solvent from the droplets, thus forming microspheres of 1 to 100µm. In case of congealing method, the molten coating material mixture is cooled or congealed to form solidified microspheres. Microspheres thus formed are separated from the spray chamber by using cyclone separator (Nidhi et al., 2016; Huang YC et al., 2003; He P et al., 1999).

3.2.2.4   Polymerization technique

This technique is generally classified into two types as (i) normal polymerization and (ii) interfacial polymerization.

Normal polymerization: Various techniques are utilized to carry out normal polymerization including bulk polymerization, suspension polymerization, and emulsion polymerization.

Bulk polymerization: This technique involves the heating of a monomer or a mixture of monomers with an initiator to start the process of polymerization. The function of initiator or catalyst is to facilitate or accelerate the rate of reaction. The drug is either adsorbed or added during the process of polymerization. The resulted polymer containing drug thus formed is fragmented or molded as microspheres (Burns P et al., 2002).

Suspension polymerization: It is carried out by heating monomers or mixture of monomers along with drugs as droplets dispersion in a continuous aqueous phase containing an initiator and other additives (Bodugoz H and Guven O, 2002).

Emulsion polymerization: Emulsion polymerization technique is similar to suspension polymerization except initiator present in the aqueous phase diffuse to the surface of the micelles or the emulsion globules in the latter stage (Liu Q et al., 2008).

Interfacial polymerization: The basic principle of this method is the reaction of monomers at the interface between the two immiscible liquid phases in order to form a polymeric film which envelops the dispersed phase. In this usually two reacting monomers are used. Out of which one monomer is dissolved and another is dispersed in the continuous phase. This method resulted in monolithic type of microspheres if the polymer is soluble in the droplets whereas reservoir or capsular form is formed if the polymer is not soluble in the droplets (Liu Q et al., 2008).

3.2.2.5   Single emulsion method

In this method, a solution or dispersion of drug and polymer is prepared in aqueous medium and the resulted mixture is then dispersed in nonaqueous medium, such as oil phase, which acts as continuous phase, leading to the formation of single emulsion. The formed emulsion is further stabilized by means of chemical crosslinking agent or heat. Depending on the types of crosslinking, the method is divided into:

Crosslinking method: This method involves the addition of specific amount of crosslinking agents in dropwise manner into already formed w/o emulsion with stirring for a particular period of time until microspheres of desired range is formed. The formed microspheres are then centrifuged, washed with suitable organic solvent and separated (Arshady R, 1989). The common crosslinker used is glutaraldehyde, formaldehyde, terephthalate chloride, epichlorohydrin and di-acid chloride (Trivedi P et al., 2008).

Thermal crosslinking method: In this method the crosslinking is carried out by dispersing aqueous phase containing drug and polymer in previously heated continuous phase under continuous stirring until a desired size range in the microspheres are formed (Dubey RR et al., 2003).

3.2.2.6   Double emulsion method

The mentioned method involves the formation of double emulsion or multiple emulsion of type w/o/w and is widely used for the encapsulation of active ingredients, including hydrophilic drugs, proteins, peptides, vaccines, enzymes etc. for their releases in controlled manner. The method starts with the formation of primary emulsion in which aqueous solution containing both drug and polymer is vigorously to form a homogenous mixture, called as w/o emulsion (Cui F et al., 2005).This primary emulsion is then sonicated prior to their addition to aqueous phase,such as solution of polyvinyl alcohol (Yan C et al., 1994), which resulted in the development of double emulsion or w/o/w. The formed emulsion is then subjected to solvent evaporation or solvent extraction with the maintenance of emulsion at reduced pressure (Ganesan P et al., 2014).

3.2.2.7   Solvent evaporation

The whole process is performed in a liquid manufacturing vehicle and used for the preparation of microspheres either by w/o or o/w or o/o types of emulsion. It involves the dispersion of polymer in a volatile solvent that is immiscible with the manufacturing vehicle phase. In this polymeric solution, core material is dispersed or dissolved and the resulted mixture is then added to liquid manufacturing vehicle phase with agitation to get desired size of microspheres. Evaporation by heating is necessary for the complete removal of volatile solvent or solvent of the polymer which leads to shrinking of polymer around the core material. Otherwise, if the core material dissolved in the polymer solution resulted in matrix type of microspheres (Ganesan P et al., 2014; Nidhi et al., 2016).

3.2.2.8　　Emulsification solvent diffusion method

This method involves two solvents, one of which is water and another is any organic solvent but must be partially or completely miscible with water, such as isopropanol, ethanol etc. In the very first step, both the solvents are mutually saturated in order to create a thermodynamic equilibrium stage. In to this, drug and polymer are added and mixed thoroughly (internal phase) and then the total mixture is emulsified with external phase (aqueous phase) using homogenization method to obtain emulsion. After the formation of emulsion, excess amount of water is added to allow the organic solvent to diffuse into continuous phase and resulting into the hardening of microspheres. In this process, there is direct addition of actives in to polymeric organic solvent and the diffusion of organic solvent varies depending on the ratio of emulsion volume to added water (typically 1:5 to 1: 10 ratio is used), solubility profile of polymer and temperature of water (Ganesan P et al., 2014; Parhi R and Suresh P, 2012).

3.2.2.9　　Ionic gelation method

In this method, anionic polymeric solution, such as alginate and chitosan are added dropwise to a solution containing cations (e.g. Ca^{2+} and Al^{3+}). This resulted in the interaction of ions present in polymer as well as in solution leading to the formation of microspheres. For rigidization, the formed microspheres were kept in original solution for 24 hr followed by filtration and drying to obtain dried microspheres.

3.2.2.10　　Phase inversion technique

Phase inversion method involves negligible loss of drug and polymer and usually having two steps: (i) the required amount of drugs is added to a polymeric solution (preferably 1-5% w/v) in suitable solvent, such as methylene chloride and (ii) the resulted mixture is then added to unstirred bath of nonsolvent (e.g., petroleum ether for methylene chloride) in a ratio (solvent to nonsolvent) of 1:100, which resulted in spontaneous formation of microspheres of desired size range. Afterward microspheres can be filtered and subsequently washed with petroleum ether and dried (Cao Y et al., 2010; Lijun D et al., 2011).

3.2.2.11   Wet inversion method

It is another method of preparation of microspheres in which acetic acid based polymeric solution added in dropwise manner into an aqueous solution containing counter ion, such as sodium tripolyphosphate. This resulted in the formation of microspheres which is then crosslinked with cross linking agents (e.g. 5% ethylene glycol diglysidyl ether) followed by washing and freeze-drying (Mi FL et al., 1999).

Microsphere preparation techniques, range of particle size, advantages and disadvantages are summarized in Table 3.1.

Table 3.1: List of methods with the size range of micro-spheres, advantages and disadvantages (Singh A et al., 2012; Nidhi et al., 2016)

Methods	Size range (µm)	Advantages	Disadvantages
Coacervation technique	1-500	• Simple and cheap method of drug encapsulation • High drug encapsulation efficiency • Efficient control over particle size with narrow size distribution, • Utilizes aqueous system for preparation, • Protection against oxidation and volatility	• Controlling pH and temperature is difficult • Aggregation of particles • Hard for scale-up in large scale production • Adjustment in reactive ratio
Hot melt microencapsulation	–	• Reproducible with respect to yield and size distribution • Suitable for the water labile polymers	• Not suitable for thermolabile substances
Spray drying and spray congealing	5-5000	• Rapid process • Good stability and retention • Formation of porous microparticles • Both hydrophilic and hydrophobic polymers can be used • Ideal for sterile product manufacturing • Complete removal of organic solvent • High encapsulation efficiency, • Heat labile material can be handled due to the use mild temperature	• Larger sized microcapsules formation due to particle agglomeration • Yield is very less due to the sticking of microparticles to the drying chamber • It can change the polymorphism of spray dried drugs • Nonuniformity in particle size and • High viscous fluids cannot be sprayed

Contd...

Contd...

Methods	Size range (µm)	Advantages	Disadvantages
Interfacial Polymerization	1-500	• High drug encapsulation efficiency • Efficient method • Fast and rapid method	• Microcapsule formed by this method are fragile and difficult to handle • Large number of washing steps are required for the complete removal of monomers and other by products, Encapsulated enzymes or protein gets inactivated due to large interface • It is difficult to control polymerization reaction.
Single emulsion method	–	• Simple and in expensive	• Cross linkers are toxic and if added in excess process became lengthy as it has to remove by series of steps, such as centrifugation, washing, separation.
Double emulsion method	–	• Offer controlled release of drug • Hydrophilic drugs, proteins and vaccines can be handled	• Coalescence and stability problem.
Solvent evaporation	0.5-1000	• Suitable for sustained delivery of lipophilic drugs as well as small molecules	• High cost of production with low drug encapsulation efficiency • Side effects due to the use of nonbiodegradable wall material or residual solvent • In case of w/o type of emulsions, removal of oil from final product is complicated • Sometimes protein may denature and formed aggregates.
Ionic gelation method	–	• Avoidance of high temperature leading to the use of thermolabile drugs • No requirements of organic solvents • Simple, cost-effective and fast method	• Relatively fast release of drug due to the low mechanical strength.
Phase inversion technique/Wet inversion method	–	• Simple and fast process There is little loss of polymer and drug	–

3.2.3   Evaluation

3.2.3.1   Particle size

Particle size is an indicator for the identification and characterization of microspheres. It is measured by coulter counter (CC) and dynamic light scattering (DLS) techniques. The principle of CC is based on the measurement of change in the electric resistance across the sensing zone when a particle in a solution of electrolyte passes through a small orifice. The resulted resistance change is then converted to particle size by a calibrated pulse-height analyzer. The measurement range of CC is in between 0.5 and 1000 μm (Parhi R and Suresh P, 2012). DLS principle is based on the measurement of the fluctuation in intensity of the scattered laser light caused by particle movement in time-dependent manner (Xu R, 2002)). The instruments based on DLS techniques are photon correlation spectroscopy (PCS) and Doppler shift spectroscopy (DSS).

3.2.3.2   Particle morphology

The surface structure or morphology of microspheres are studied by various microscopic techniques including light microscopy (LM), polarized light microscopy (PLM), fluorescence microscopy (FM), confocal microscopy (CM), scanning electron microscopy (SEM), transmission electron microscopy (TEM) and atomic force microscopy (AFM). Out of those, SEM, TEM and AFM are widely used in recent times. These techniques give detail information of particle morphology along with their size, size distribution and internal structure. In SEM technique, the bombardment of sample with a focused electron beam that in turn generates secondary electrons from the specimen surface. Then, the surface characteristic is obtained from these emitted secondary electrons (Domingo C and Saurina J, 2012). The principle of TEM technique is based on the interaction of transmitted beam of electron with an ultrathin specimen during its transmission. This lead to the formation of 2D images on an imaging device. Resolution power of TEM is higher than that of SEM and can generate internal structure of the microspheres (Klang V et al., 2012). The principle involved in AFM instrument is the interaction force between the tip of tiny probe (mounted over cantilever) and the sample. The changes in the movement of probe in sample are converted into electrical signal employing laser detection systems. Like SEM, 3D images are formed with AFM, but with higher resolution of up to 0.01 nm (Mehnert W and Mäder K, 2012). Beside above, there are few modified techniques, such as environmental SEM (ESEM), cryo-SEM and -TEM, and freeze fractures SEM and TEM (FF-SEM and FF-TEM) are also used measure size as well as surface morphology.

3.2.3.3   Entrapment efficiency and drug loading

Entrapment efficiency (EE) and drug loading (DL) of microspheres are generally measured in three steps: (i) separation of microspheres from liquid medium using methods, such as ultrafiltration, ultracentrifugation or size exclusion chromatography, (ii) measurement of free/dissolved drug in the liquid medium by spectroscopic or chromatographic method, and (iii) deduction of this free/dissolved drug from the initial amount of drug, the amount of drug encapsulated can be obtained. From the above, EE and DL can be calculated using following equations (Neupane YR et al., 2014):

$$EE (\%) = \frac{W_{\text{initial durg}} - W_{\text{free drug}}}{W_{\text{initial drug}}} \times 100$$

$$DL (\%) = \frac{W_{\text{initial durg}} - W_{\text{free drug}}}{W_{\text{polymer}}} \times 100$$

where winitialdrug and wfreedrug are the mass of initial drug taken for the incorporation into microsphere and the mass of the drug detected in the supernatant liquid, respectively.

3.2.3.4   Isoelectric point
The electrophoretic mobility of microspheres is measured using apparatus microelectrophoresis. Then, isoelectric point can be determined from this electrophoretic mobility data. The electrical mobility is determined by measuring the mean velocity of particles over a distance of 1 mm at different pH values ranging from 3 to 10. This electrophoretic mobility can be related to ionizable behaviour or ion absorption and surface containing charges of microspheres (Ganesan P et al., 2014).

3.2.3.5   Density measurement
Multivolume pychnometer is used to determine density of the microspheres. In the first step, accurately weighed sample in a cup is placed inside the equipment; this is followed by the introduction of helium in to the chamber. Helium is allowed to expand which results in the reduction in pressure in the chamber. From the data obtained from the two consecutive readings at different initial pressure and volume, and hence the density of the microspheres is calculated (Sinha VR et al., 2005).

3.2.3.6   Angle of contact
The wetting property is determined by measuring angle of contact of the microspheres. The angle of contact is usually measured at either the solid/air/water interface by placing a droplet in a circular cell of inverted microscope. Both advancing and receding angle of contact are measured at 20°C within a minute of deposition of microspheres (Kawashima Y et al., 1991).

3.2.3.7   In vitro methods

In vitro drug release from the microspheres was carried out employing four types of arrangements, such as (i) dialysis bag method, (ii) modified Franz diffusion cell, (iii) interface diffusion systems, and (iv) dissolution apparatus.

(i) Dialysis bag method

The apparatus constitutes of a dialysis bag and a glass tubes (Figure 3A). The sample microparticles have to place in dialysis bag and then the bag is dipped in a screw capped glass tubes containing appropriate medium. The whole apparatus is reciprocated in a horizontal shaker and samples can be withdrawn at desired time interval (Moebus K et al., 2009).

(ii) Modified Franz diffusion cell

In this diffusion cell a regenerated cellulose filter membrane and thin cloth are attached tightly to the lower end of a polypropylene tube, which is placed vertically in a plastic vessel filled with desired medium (Figure 3B). The tube is placed in such that the filter is only wetted, but not submerged in the medium. The microparticles are placed on the filter membrane and then sealed with paraffin to avoid evaporation of solvent (Moebus K et al., 2009).

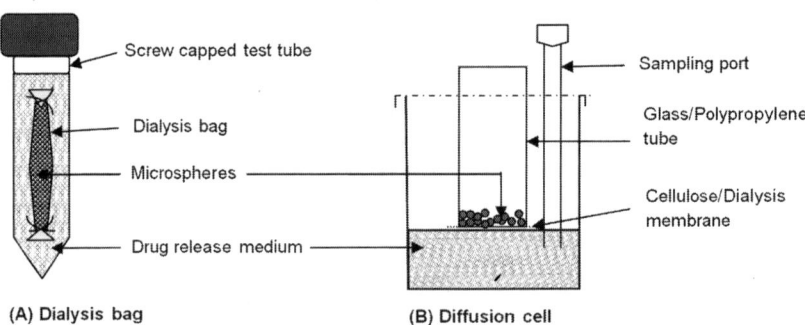

Figure 3.3: Schematic illustration of experimental setups used for *in vitro* drug release determination: (A) dialysis bag and (B) diffusion cell

(iii) Interface diffusion system

This system consists of four compartments. Compartment A simulates the oral cavity and an appropriate concentration of drug in a buffer is placed initially. Buccal membrane is represented by compartment B containing 1-octanol. The compartment C simulates stomach and having 0.2M HCl and compartment D representing protein binding containing 1-octanol. Before use, the aqueous phase and 1-octanol were saturated with each other, and samples were withdrawn at desired time period with the help of a syringe (Ganesan P et al., 2014).

(iv) Dissolution apparatus

In vitro release profiles of diverse types of microspheres can also be carried out using rotating paddle and basket apparatus (United States Pharmacopeia, USP/British Pharmacopeia, BP) (Parodi B et al., 1996; Cassidy JP et al., 1993). The dissolution medium used for the study ranges from 100 to 500 mL and speed of rotation of rotating elements varied from 50 to 100 rpm. An equivalent amount of microspheres are placed in the apparatus containing appropriate medium and the samples are collected at different time intervals. The same amount of medium is replaced after each sampling and the samples are analyzed as per monograph requirements. Thereafter, the release profile of drug is determined by plotting the release against time.

3.2.3.8   In vivo methods

Among various *in vivo* methods, the most reliable and widely used methods are buccal absorption tests and animal models.

(i) Buccal absorption test

This test is based on the measurement of the extent of drug loss from the human oral cavity after oral administration of single and multi-component drug system. Apart from that the test was used successfully to determine the influence of drug structure, contact time, pH of the oral fluid and initial drug concentration during the holding period of drug system in the mouth cavity.

(ii) Animal models

Till date, a number of animal models reported in the literature including rats, rabbits, cats, dogs, hamsters, pigs and sheep (Ganesan P et al., 2014). These animal models are used for the screening of various microparticles containing diverse class of drugs and also to find out mechanisms and efficiency of various penetration enhancers. The common procedure is used for all animal models involving anesthetizing the animal followed by administration of microparticles and at different time points, the blood samples are withdrawn and analyzed for the drug concentration. In addition, pharmacological action also can be measured after administration of microspheres containing drug through different routes.

In vitro–in vivo correlations

It is insufficient to predict the therapeutic efficacy using only *in vitro* dissolution study. Therefore, an *in vitro* dissolution and *in vivo* bioavailability not only ensure batch to batch consistency (reproducibility) but also to serve as a tool in the development of new dosage forms with desired *in vivo* performance. This correlation can be established on the basis of plasma level data, urinary excretion data, and pharmacological response.

3.2.4   Applications

There are many applications of microspheres out of which few important one is described below:

3.2.4.1   Drug delivery in various routes

Microspheres developed with polymers with desired properties demonstrate favourable biological behaviour, such as bioadhesion, mucoadhesion, small size with suitable physicochemical characteristics which make it unique to be used in various routes such oral, ocular, topical, transdermal, nasal, buccal, colon, and vaginal routes (Sahil K et al., 2011; Saralidze K et al., 2010; Lokwani P et al., 2011; Dhakar R C et al., 2010; Lin CY et al., 2015; Hafeli U, 2002; Khan MS and Doharey V, 2014).

3.2.4.2   Gene and vaccine delivery

Microspheres can be used in the gene delivery in the oral route due to adhesive properties of certain bioadhesive polymers, such as chitosan, gelatin, etc. Several vaccines, such as tetanus, diphtheria vaccines are incorporated in microspheres made up of biodegradable polymer can administered in parenteral routes thereby avoiding problems associated with conventional vaccines (Virmani T and Gupta J, 2017).

3.2.4.3   Delivery of monoclonal antibodies

Monoclonal antibodies are highly specific molecules which bind to specific antigen present in cell or tissue of the body and showing its therapeutic or diagnostic action. Furthermore, monoclonal antibodies can be adhered to the surface of microsphere resulting in the drug release in the desired site of the body.

3.2.4.4   Drug targeting

Drug targeting can be achieved via passive as well as active targeting method. In passive method, microspheres without surface modifications are used to deliver the drug in the desired site. Lack of specificity may lead to a small fraction of administered drug reached to desired site. On the other hand, the surface of microspheres is modified with the attachment of antibodies, enzymes, polysaccharides and protein to enhance the specificity towards desired site which helps the delivery system to deliver maximum amount of drug in the desired site. This not only reduces the side effect caused due to distribution of drug (especially cytotoxic drugs) to normal cells, but also decreases the dose of the drug (Roser M et al., 1998).

3.2.4.5 Decrease toxicity and increase efficacy

Microspheres as drug delivery system capable of decreasing toxicity and increasing efficacy by delivering the drug in desired site, controlling the drug absorption and biodistribution and also it can prevent drug interaction in case of multidrug resistance by physically separate the drugs (Vallelado AI et al., 2002).

3.2.4.6 Protection of active ingredients

Microspheres can protect the encapsulated drugs from harsh environment of stomach by delaying the drug release. This is possible when microspheres are either coated with enteric polymer or the matrix is made up of enteric polymer (Sankavarapu V and Aukunuru J, 2009).

3.2.4.7 Diagnostic application

Specifically radioactive microspheres can be used for the imaging of various body parts, such as liver and spleen for tumours, and also can be used for local radiotherapy, local restenosis in coronary artery. Diameter of microspheres plays a vital role in the target imaging site. For instance, radiolabelled microspheres are injected via IV route and portal vein generally entrapped in the lungs. Therefore, this route can be considered for the scintigraphic imaging of lung tumour cells (Ganesan P et al., 2014).

A list of commercially available microsphere formulations with their manufacturer, active ingredients, use and delivery route are presented Table 3.2.

Table 3.2: Commercially available microsphere formulations for application in different routes (Singh A et al., 2012)

Commercial name	Manufacturer	Drug	Indication	Delivery route
Nopap™	Mayne Pharma International	Paracetamol	Pain and fever	Oral
Micro-K® Extencap®	KV Pharmaceutical Co.	Potassium chloride	Hypokalaemia	Oral
CIPRO®	Bayer Inc.	Ciprofloxacin	Microbial infection	Oral
Cardioviva™	Micropharma Ltd.	Lactobacillus reuteri NCIMB 30242	Hypercholesterolaemia	Oral

Contd...

Contd...

Commercial name	Manufacturer	Drug	Indication	Delivery route
Norvir®	Abbott Laboratories	Ritonavir	HIV	Oral
Sandimmun Neoral®	Novartis Int. AG	Cyclosporin A	HIV	Oral
Arestin®	Orapharma	Minocycline	Periodontitis	Subgingival
Trioxil®	DS Laboratories	Ginseng and hamamelis extracts, arnica	*Propionbacterium acnes* bacteria and fungi	Transdermal
Tagravit™	Tagra Biotechnologies Ltd.	Vitamin A, E, F	Aging	Transdermal
Nutropin®Depot	Genentech/ Alkemes	Somatropin	Growth deficiencies	S.C.
Zoladex®	I.C.I.	Goserelin acetate	Prostate cancer	S.C.
Vivitrol®	Alkemes	Naltrexone	Alcohol dependence	I.M.
RISPERDAL® CONSTA®	Janssen/ Alkemes	Risperidone	Schizophrenia;bipolar disorder	I.M.
Trelstar™ depot Decapeptyl®SR	Pfizer Ferring	Triptorelin	Prostate cancer	I.M.
Somatuline®LA	Ipsen-Beafour	Lanreotide	Acromegaly	I.M.
Suprecur®MP	Sanofi-Aventis	Buserelin	Endometriosis	I.M.
Enantone Depot® Trenantone® Enantone Gyn Lupron Depot®	Takeda Takeda Takeda TAP	Leuprolide	Prostate cancer/ endometriosis	I.M.
Sandostatin®LAR	Novartis	Octreotide	Acromegaly	I.V. or S.C.

* S.C. = subcutaneous, I.M. = intramuscular, I.V. = intravenous

3.3   Monoclonal Antibodies

Antibodies, also called as immunoglobulin's (Ig), are chemically glycoproteins naturally formed in response to invading antigens (foreign particles) including microorganisms and viruses. Therefore, they played a vital role in developing immunity or internal defiance against various diseases and infections. This is possible as the antibodies have the ability to recognize and bound to specific antigen and the region of an antigen which interacts with the antibody is called as epitope. Large group of antibodies that are formed from the human/ animal body can recognize many epitopes of a particular type of antigen. Among five common isotypes (IgG, IgE, IgA, IgM and IgD), IgG has most therapeutic potential because it has longer half-life and capable of permeating extravascular spaces when compared with other isotypes. These antibodies are produced from many antibody producing plasma cells; therefore, called as polyclonal (Rauta PR et al., 2012; Roopenian DC and Akilesh S, 2007; Schlachetzki F et al., 2002; Wang J et al., 2015; Loisel S et al., 2007; Hérodin F et al., 2005; Martin PL and Weinbauer GF, 2010). In fact, this leads to better fighting against infections and provides higher degree of immunity. However, heterogeneity of these antibodies is a constraint for their use as research tool.

Monoclonal antibodies (mAbs) are highly specific and dominant group of recombinant proteins used in human therapy. These antibodies were first successfully developed by Kohler and Milstein in their laboratory in 1975. They removed B cells (plasma cell precursors) of mice spleen that has been immunized with a predetermined antigen, and then fused with an unrestricted growth of myeloma cells *in vitro*. This lead to the formation of single-cell hybrid called as hybridoma. In this, the B cells confer antibody production potential, whereas myeloma cells enable the single cell to grow well in culture and divide indefinitely. These hybridomas have the potential to secret only one antibody for long period selective for a single epitope called as monoclonals. The innovation of mAbs technology not only brings revolution in the field of immunology in the biomedical field but also used in all medical field be it diagnostic or treatments of various diseases.

3.3.1   Structure of Monoclonal Antibodies

Molecular weight of a typical IgG is approximately 150 kDa and is a *Y*-shaped molecules composed of four polypeptide subunits with two identical heavy chains (each of which has approximately 50 kDa) and two identical light chains (each of which has approximately 25 kDa) (Figure 3.4). The N-terminus of each heavy chains associates with one of the light chains to form two antigen binding domain. These are called as fragment antigen binding (Fab) domains. The C-terminus is having two heavy chains, which combine to form the fragment of crystallization (Fc) domain. Therefore, the N-terminus and C-terminus are termed as arms and tails of antibodies, respectively. The Fab domain provides structural basis for the antigen and is responsible for antigen binding/reorganization. On the other hand, Fc domain is important in antibody interaction with effector cells, such as macrophages and for activation of complement cascade. The heavy chain is composed of three constant domains and one variable domain (each domain consisting of 110 amino acids), whereas light chain contains one each of constant and variable domain. Further the variable domain in both heavy and light chain consisted of three small sections of peptide termed as complementarity determining regions (CDRs). CDRs are hypervariable regions which determine antibody binding specificity (Saeed AFUH and Awan SA, 2016).

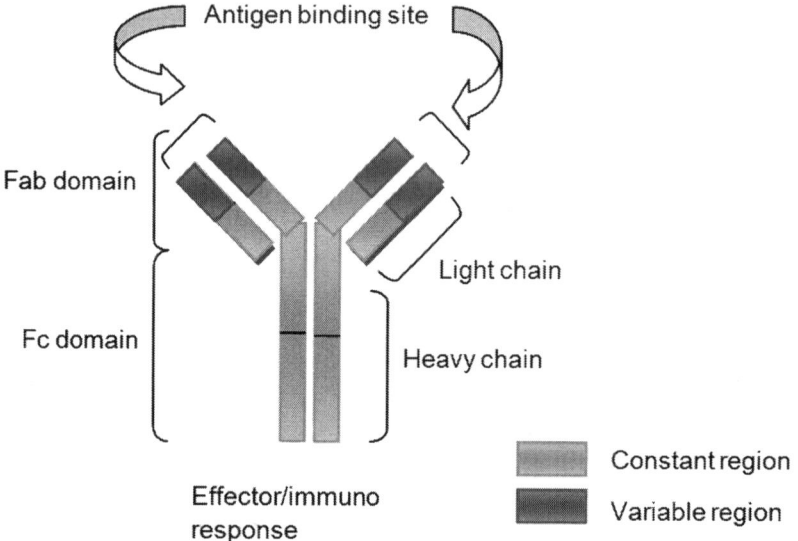

Figure 3.4: Structure of monoclonal antibody and its components

3.3.2   Progress in the Development of Monoclonal Antibody

Although the first mAb developed from murine, there are another three stages of development of mAb, such as chimeric antibodies, humanized antibodies and human antibodies (Gonzales NR et al., 2005). The main focus point of the above progress in antibody development was reducing their immunogenicity.

3.3.2.1   Murine monoclonal antibodies

In this both constant and variable region were derived from the mouse. These antibodies have disadvantages, such as immunogenicity (produces human antimouse antibodies) resulted in the reduced clinical efficacy, weak affinity to Fc receptor on the surface of immune cells, short duration of action, poor killing ability and sometimes may induce allergic reaction (Han L et al., 2013).

3.3.2.2   Chimeric antibodies

It is developed by replacing constant region of murine mAbs with human sequence. This replacement is carried out by gene recombination technique. Thus, chimeric mAbs retain its antigen binding potential and at the same time it reduces immunogenicity to minimum level. However, it is unable to completely remove the immunogenicity problem because the antibodies still have 30% of murine antigen.

3.3.2.3   Humanized antibodies

These are prepared by replacing all the sequence of murine mAbs with human except CDRs (hypervariable region). As these reconstructed antibodies still retain remaining amount of heterologous genes of murine, it may cause immune rejection. The other disadvantages are low specificity and low affinity. Surficial remodeling method is used to replace or rehabilitate amino acid residues that are different from humans to reduce heterology without disturbing activity of antibodies (Roopenian DC and Akilesh S, 2007; Lin Y and Yan XY, 2004).

3.3.2.4   Human antibodies

This is fully human mAbs prepared by replacing total rodent sequence by human sequence and considered as most ideal as the modification reduces the possibility of immunogenicity and treatment failure (Singh S et al., 2018).

3.3.3   Preparation of Monoclonal Antibody

3.3.3.1   Hybridoma technology

Hybridoma technology is used to develop stable cell lines which secret a defined mAb for considerable period of time. To achieve this, B-cells from an immunized animal are fused with a myeloma cell line in the presence of fusing agent, polypropylene glycol (PEG) to form immortalized hybridoma cells. The resulted cells are cloned by limiting dilution and screened to identify specific clones producing identical antibodies called as monoclonal antibodies. These cloned antibodies are then expanded in culture to generate high quantity of desired mAb (Kohler G and Milstein C, 1975). Hybridoma technology was first applied to generate murine (mouse) mAbs in 1975. Since the inception of technology, thousands of mAbs against a wide variety of antigens were developed. All are classified into two hybridoma types: hetero-hybridoma and homo-hybridoma. Hetero-hybridomas are the antibodies generated from host B cells and fusion cells line derived from two different species. Such type of hybridoma cells are first developed in 1988 with B-cell from mouse and fusion cell line from rabbit (Beerli RR and Rader C, 2010; Raybould TJ and Takahashi M, 1988). Hetero-hybridomas are unstable, comparatively inefficient and incapable to secret antibodies for prolonged period of time. In case of homo-hybridoma, both the B-cell and fusion cell line are obtained from same species, such as rabbit–rabbit hybridomas (Beerli RR and Rader C,

2010; Spieker-Polet H et al., 1995).Rabbit–rabbit homo-hybridomas are also less stable when compared with mouse hybridomas. The low efficiency of cell fusion is the main limitation of this method.

3.3.3.2   Phase display technique

Bacteriophase or simply phase is viruses that have the ability to infect and replicate within bacteria. Phase display technology was developed by G. Smith 1985. According to this technique, the harvested variable genes from lymphocytes and then the combination of heavy chain variable (VHs) and light chain variable (VLs) are cloned and expressed on the surface of filamentous bacteriophage by fusion to its minor coat protein (pIII) without affecting infectivity of phage. This is followed by the selection of phase particles bearing expressed specific mAbs on their tips (Saeed AFUH and Awan SA, 2016; Weber J et al., 2017). In this process, these phase particles have the genetic information encoding of the modified coat protein and therefore physically linking genotype (single stranded DNA of the specific virion) and phenotype (phase coat fusion protein expression). The hybridoma technique is generally confined to rodents, whereas phase display can be successfully employed on any species whose Ig genes are identified or known to generate desired mAbs (Lowe D and Jermutus L, 2004; Peterson NC, 2005). The main limitation of this technique is that the larger protein or peptide sequence can pose challenges due to reduction in infectivity. In order to maintain the infectivity, an additional helper phage that helps the phase genome to encode all proteins necessary for harvesting infectious phase particles. This technique is so popular that a first filamentous phage display antibody libraries based on pIII fusion protein are published in early 1990s (Clackson T et al., 1991; Barbas CF et al., 1991). In 2000, the total process of phage display technology was described in selecting the rabbit mAbs (Rader C et al., 2000). Presently, phase display technique is being used to generate complete human antibodies for various therapeutic purposes because of its ability to control various desired parameters along with ease and high speed antibody generation (Shim H, 2016). The loss of the natural cognate pairing of heavy chain and light is the main disadvantage of display technique (Tiller T, 2011). Furthermore, this technique is dependent on the number of factors including the construction of phage vectors, helper phage, host cells, phage library and bioplanning processes (Ribatti D, 2014).

3.3.3.3   Single B cell amplification

To circumvent the limitations of hybridoma and phase display techniques, single B cell antibody technology has been developed. This technique is based on the direct cloning of identical pairs of antibody VH, light chain lambda variable (VLλ) and light chain kappa variables (VLκ) genes obtained from an antigen-specific memory B cells or plasma/plasmablast cells (ASPCs) by using polymerase chain reaction (Love JC et al., 2006; Wrammert J et al., 2008; Rawstron AC, 2006). The total technology is having following six small steps (Zhang Z et al., 2017). Step 1: Identification and isolation of specific single B cell. This is carried out by either random way or fluorescence-activated cell sorting (FACS) method from lymphoid tissues or peripheral blood.

Step 2: Reverse-cell transcription-polymerase chain reaction (RT-PCR) on the isolated single cell with antibody-specific primers.

Step 3: PCR mediated amplification of Ig genes and sequencing.

Step 4: Ig genes are cloned into expression vector.

Step 5: The expression of Ig genes in bacteria, such as *Escherichia coli* or mammalian cell systems, such as CHO cells and HEK 293.

Step 6: Purification and evaluation of proteins with enzyme-linked immunosorbent assay.

This method has advantages including the retention of cognate pairing of light and heavy chain that provide platform to exploit antibody affinity and specificity in natural manner resulting into generation of mAbs with all desired characteristics, such as specificity, affinity and stability profile. In addition, this technique can be used to generate mAbs from mice, rabbits and human and those mAbs can be used to treat various diseases, such as autoimmune disorder, infectious disease and cancers (Tiller T, 2011). Like other method, this method also suffers, such as the requirements of expensive equipments and technical skills, particularly to isolate ASPCs and then removal of VH and VL genes.

3.3.4   Application

Muromonab (OKT3) is the first mAbs to be approved by Food and drug administration (FDA) for the treatment in late 1980s. After that, during the course of three decades there is large number of mAbs approved and used for not only treatment but also for diagnostic purposes. Therapeutic mAbs are used to treat cancers, autoimmune diseases, miscellaneous diseases, including chronic gastrointestinal conditions (e.g. Crohn's disease), and respiratory conditions, such as asthma and central nervous system degenerative disorders (e.g. multiple sclerosis).

3.3.4.1   Therapeutic use
3.3.4.1.1   For the treatment of cancers
Rituximab (Rituxan) was the first mAb to be approved by FDA in 1997 for the treatment of Non-Hodgkin's Lymphoma (NHL), including marginal

lymphoma, chronic lymphocytic leukaemia, and mantle cell lymphoma (Maloney DG et al., 1994). It is a chimeric mAbs (mouse-human) against the target cluster differentiation (CD20) expressed on the surface of normal B cells, on >90% of B-cell neoplasms (Ribatti D, 2014).The mechanism of action of rituximab is complement-dependent cytotoxicity, antibody-dependent cytotoxicity and sensitization of tumour cells by drugs or radiation (Eisenbeis CF et al., 2003). After infusion of rituximab, the normal B-cell count decreases to zero but complete recovery take place by 9-12 months.

Cetoximab (Erbitux), a chimeric IgG1, was approved in 2004 for the treatment of colorectal carcinoma. This mAbs binds to human epidermal growth factor (HER-1), a member of epidermal growth factor receptor (EGFR), and act by inhibiting cell growth, decrease vascular endothelial growth factor and induction of apoptosis (Galizia G et al., 2007; Kawaguchi Y et al., 2007).

The overexpression of Her-2 is usually seen in case of breast cancer cells. To treat this, a humanized mAb trastuzumab (Herceptin) was approved. It acts by binding with Her-2 receptor and inducing expression of antiangiogenic and suppressing pro-angiogenic factors, and cytotoxicity mediated by antibody (Hudis CA, 2007; Sledge Jr GW, 2004).

Another humanized IgG4 mAb gemtuzmab ozogamicin (Mylotarg) was approved in 2000 for the treatment of patients suffering from acute myeloid leukaemia. Its target receptor is CD-33 which is expressed on most of the leukemic progenitor cells, myeloid leukaemia blasts and normal and myelo-monocytic haematopoietic progenitor cells conjugated to a cytotoxic antibiotic, calicheamycin (Pagano L et al., 2007).

In 2001, Alemtuzumab (Campath), a humanized IgG1 mAb was approved by FDA for the treatment of drug-resistant chronic lymphocytic leukaemia (Alinari L et al., 2007). They bind to cell surface glycoprotein CD-52 present on the T and B lymphocytes leading to tumour cell death through complement-dependent cytotoxicity and antibody dependent cytotoxicity. It has been used in case of haematologic malignancies, where it induces the cell death generated from allogenic transplantation grafts (Naparstek E et al., 1999).

In 2006, the first fully human mAbs panitumunab (Vectibix) was approved by FDA. It was developed by transgenic technology and acts by blocking the action of EGFR. Panitumunab is being used for the treatment of colorectal cancer (Jakobovits A et al., 2007).

Ipilimumab is another fully human mAbs approved by FDA. Ipilimumab is IgG type of mAb specific for cytotoxic T lymphocyte antigen-4 (CTLA-4) present on the surface of helper T cells. It acts by inhibiting the development of peripheral immune tolerance and presently used for the treatment of melanoma (Acharya UH and Jeter JM, 2013).

An antibody–drug conjugate (ADC), brentuximab vedotin with proven efficacy in patient having CD30+ antigen malignancies, such as classical Hodgin lymphoma (HL) and systemic anaplastic large cell lymphoma patients (Singh S et al., 2018).

FDA has approved the first mAb having two binding sites (bispecific), Catumaxomab for the treatment of malignant ascites. This is chimeric (mouse and rats) type of mAb with specificity towards CD3 and epithelial cell adhesion molecules (Linke R et al., 2010).

3.3.4.1.2   Autoimmune disease

In 1998, FDA approved infliximab for the treatment of autoimmune disease including ankylosing spondylitis (AS) rheumatoid arthritis (RA), ulcerative colitis (UC), plaque psoriasis, psoriatic arthritis (PA) and Crohn's disease (CD). Infliximab has the affinity to human tumour necrosis factor (TNFα) and inhibits its soluble as well as transmembrane bioactive form (Singh S et al., 2018). Adalimumab was approved by FDA in 2002 and has the same mechanism and application (Singh S et al., 2018).

Ustekinumab is another mAb approved by FDA for the treatment of human plaque psoriasis, CD and PA. The p40 protein subunit of interleukin-12 (IL-12) and IL-23 is responsible for inflammatory and immune responses. Ustekinumab binds to p40 protein subunit and stops both inflammatory and immune responses (Singh S et al., 2018).

Belimumab is FDA approved fully human mAb that blocks the binding of soluble B lymphocyte stimulator to its receptor, thereby inhibits the survival of B cells. It is used for the treatment systemic lupus erythrematous (SLE) since 2011 (Singh S et al., 2018).

Daclizumab, a humanized mAb, has the affinity towards CD25, a subunit of the high-affinity IL-2 receptor. Daclizumab is used in the treatment of multiple sclerosis where it supposed to involve in the modulation of IL-2 mediated activation of lymphocytes (Singh S et al., 2018).

3.3.4.1.3   Organ transplant rejection

Muromonab is the first FDA approved murine mAb for the prevention of kidney transplant rejection. Later on, it is being used to treat all types of organ transplant rejection (acute graft versus disease) and act probably by blocking T cells reverse graft rejection (Rodgers KR and Chou RC, 2016).

Another mAb used for transplant rejection (prophylaxis of renal transplant rejection) is basiliximab. It is of chimeric type mAb and acts by inhibiting IL-2 mediated activation of lymphocytes through competitively and specifically binding to IL-2Rα (Ribatti D, 2014).

3.3.4.1.4   Heart diseases

The first chimeric therapeutic mAb, abciximab, was approved by FDA for the pre-surgical prevention of thrombosis for coronary artery interventions (Lefkovits J and Topol EJ, 1995). It prevents the binding of fibrongen, von Willebrand factor and other activating molecules to glycoprotein IIb/IIIa receptors on platelets, thereby inhibiting platelet aggregation (Accessdata.fda 2018).

Idarucizumab, a humanized mAb, has higher binding affinity to dabigatran and its acyl glucuronide metabolites resulting in neutralization of anticoagulant effect of dabigatran by preventing it from binding with thrombin.

Alirocumab and evolocumab are fully human mAbs and proprotein convertase subtilism/Kexin type 9 (PCSK-9) is their target. They bind to PCSK-9 and inhibit the binding of PCSK-9 to low density lipoprotein (LDL) receptor leading to the number of LDL receptors available to metabolse LDL, thus lowering of LDC-C level in the body. Alirocumab is used to treat hypercholesterolaemia, whereas evolocumab employed for primary hyperlipidaemia and homozygous familial hypercholesterolaemia.

3.3.4.1.5   Respiratory diseases

In 2003, FDA approved humanized mAb omalizumab for the treatment of allergic asthma and chronic idiopathic urticaria. The mechanism of action of omalizumab is to inhibit the binding of IgE to the high-affinity crystallizable epsilon R1, an IgE receptor fragment, on the basophils and mast cells surface. This results in the decrease of mediators of allegic response.

For the treatment of asthma, currently there were two humanized mAb, such as mepolizumab and reslizumabis available in the market. They supposed to inhibit the signaling of IL5 that resulted in the production and survival of eosinophils.

Another mAb dupilumab (fully human) is acts by blocking IL-4 and IL-13 and used in the treatment of atopic dermatitis.

3.3.4.1.6   Eye diseases

Eye diseases, such as macular edema, diabetic macular edema, neovascular (wet) age related macular degeneration, diabetic retinopathy with DME are treated with humanized mAb ranibizumab. It binds to receptor present on the vascular endothelial growth factor A (VEGF-A) of endothelial cells, thereby reducing endothelial cell proliferation, vascular leakage and new blood vessel formation.

3.3.4.1.7   Kidney disease

Presently, only one approved mAb eculizumab is being used to treat kideney elements, such as haemolytic uraemia syndrome and paroxysmal nocturnal haemoglobinuria. The mechanism of action of eculizumab is to inhibit the cleavage of complement protein C5 to C5a and C5b, also the generation of the terminal complement complex C5b-9.

3.3.4.1.8   Infection

In 1998, a humanized mAb palivizumab was approved for use in the prophylaxix of respiratory synctial virus (RSV) in children. The mechanism of action palivizumab is to neutralize and fusion–inhibition activity of RSV by binding with its antigenic site of the F protein.

Both obiltoxaximab (a chimeric type mAb) and raxibacumab (a fully human type mAb) used to treat inhalation anthrax. They act by binding with protective antigen component of *Bacillus anthracis* toxin.

Bezlotoxumab is another fully human IgG1 based mAb approved in 2016 for the treatment of *Clostrodium difficile* infection by binding and neutralizing its toxin B.

3.3.4.2   Diagnostics

Technetum (99mTc)-arcitumomab (CEA-Scan) is an immunoconjugate approved by FDA in 1999 for the diagnostic imaging of colorectal cancers. It is composed of Fab' fragments of murine and IgG1 mAb generated from ascites of mice (Fiorella G et al., 2001).

111In-capromab pendetide is a radioisotope conjugate of radionuclide indium-111 and capromab, a mAb obtained from mouse. This is linked to pendetide to chelate the radionuclide. It is marketed in the trade name of prostascint, which capable of recognizing a protein found on both normal and prostate cancer cells. Single photon emission computed tomography is employed for imaging following intravenous injection (Manyak MJ, 2008).

In 1996, the trade name Verluma containing 99mTc-nofetumonab-merpenane was approved by FDA for the diagnosis of lung cancer and carcinoma in breast, gastrointestinal tract, ovary, pancrease, cervix, kidney and bladder. It is consisting of mAb part, nofetumomab that linked to chelator merpentane. This conjugate is then linked to radioisotope technetium-99m. The Fab part of the antibody is generated from murine mAb, which is able to recognize the pancarcinoma glycoprotein antigen (Straka MR et al., 2000; Breitz HB et al., 1997).

Yttrium-90 (90Y)–ibritumomab tiuxetan (Zevalin) is consisting of mAb ibritumonab and chelator tiuxetan along with radioactive isotope Yttrium-90.

It was approved by FDA in 2002 and acts by binding of mAb part with CD20 antigen present on the surface of normal and malignant B cells. This resulted in the death of nearby cells, mostly by beta emission from the radioisotopes. In addition, mAb itself trigger cell death through antibody-dependent cellular cytotoxicity (ADCC), complement dependent cytotoxicity (CDC) and apoptosis. Thus, it is used as radioimmunotherapy treatment of relapsed or refractory, transformed B-cell non-Hodgkin's and a lymphoproliferative disorders (Milenic DE et al., 2004). The combination of tositumomab, a murine IgG2a lambda mAb, with radioisotope iodine 131 formed 131I-tositumomab. This is approved for the treatment of refractory non-Hodgkin lymphoma (Srinivasan A and Mukherji SK, 2011).

Despite of wide application, the treatment of various diseases with mAbs is limited to only few developed countries because of high production cost, unclear mode of action and associated limitations, and the unavailable data regarding their pharmacokinetics (Chames P et al., 2009). Table 3.3 presented a list of approved antibodies with their composition, trade name and uses.

Table 3.3: Various antibody classes with their origin, approved antibodies, trade name and uses (Rodgers KR and Chou RC, 2016)

Antibody classification	Make	Approved antibodies	Trade name	Use
Murine	Fully murine antibody	Muromonab-CD3	Orthoclone OKT3	Kidney transplant rejection
Chimeric	Mouse variables regions with human constant regions	Rituximab	Rituxan	Non-Hodgkin's Lymphoma (NHL), including marginal lymphoma, chronic lymphocytic leukaemia, and mantle cell lymphoma
		Abciximab	ReoPro	Pre-surgical prevention of thrombosis for coronary artery interventions
		Cetuximab	Erbitux	Colorectal carcinoma
		Infliximab	Remicade	Ankylosing spondylitis (AS) rheumatoid arthritis (RA), ulcerative colitis (UC), plaque psoriasis, psoriatic arthritis (PA) and Crohn's disease (CD).
		Basiliximab	Simulect	Prophylaxis of renal transplant rejection
		Obiltoxaximab	Anthim	Inhalation anthrax

Contd...

Contd...

Antibody classification	Make	Approved antibodies	Trade name	Use
Humanized	Murine CDR, human Fc and framework regions	Palivizumab	Synagis	Prophylaxis of respiratory syncytial virus (RSV)
		Idarucizumab	Praxbind	Thrombosis
		Trastuzumab	Herceptin	breast cancer
		Gemtuzmab ozogamicin	Mylotarg	acute myeloid leukaemia
		Alemtuzumab	Campath	Drug-resistant chronic lymphocytic leukaemia
		Daclizumab	Zenapax	multiple sclerosis
		Omalizumab	Xolair	Allergic asthma and chronic idiopathic urticaria
		Mepolizumab	Nucala	Asthma
		Reslizumabis	Cinqair	Asthma
		Ranibizumab	Lucentis	Eye diseases, such as macular oedema, diabetic macular oedema, neovascular (wet) age related macular degeneration, diabetic retinopathy
		Eculizumab	Soliris	Kidney elements, such as haemolytic uraemia syndrome and paroxysmal nocturnal haemoglobinuria
Human	Fully human antibody	Adalimumab	Humira	AS, RA, UC, PA, and CD
		panitumunab	Vectibix	Colorectal cancer
		Ipilimumab	Yervoy	Melanoma
		Ustekinumab	Stelara	Plaque psoriasis, CD and PA
		Belimumab	Benlysta	Systemic Lupus Erythrematous (SLE)
		Alirocumab	Praluent	Hypercholesterolaemia
		Evolocumab	Repatha	Primary hyperlipidaemia and homozygous familial hypercholesterolaemia
		Raxibacumab	Perjeta	inhalation anthrax
Fragments	Fab	Ranibizumab (Humanized)	Lucentis	Macular degeneration
Conjugate	Radiolabeled toxin conjugated	99mTc-arcitumomab (Murine)	CEA-Scan	Colorectal cancer
		111In-capromab pendetide (Murine)	Prostascint	Prostate cancer
		99mTc-nofetumonab-merpenane (Murine)	Verluma	Lung cancer and carcinoma in breast, gastrointestinal tract, ovary, pancrease, cervix, kidney and bladder
		90Y-ibritumomab tiuxetan (Murine)	Zevalin	B-cell non-Hodgkin's and a lymphoproliferative disorders
	Drug conjugates	Brentuximab vedotin (Chimeric)	Adcetris	Classical Hodgin Lymphoma (HL) and systemic anaplastic large cell lymphoma

3.4   Niosomes

Conventional dosage forms are associated with various disadvantages including low bioavaiability, poor aqueous solubility, low membrane permeability, poor patient compliance, variable plasma drug concentration and undesirable effects in some cases (Bochot A and Fattal E, 2012; Ahmad MZ et al., 2014; Song S et al., 2015). Since last few decades, with various novel drug delivery approaches including nanoparticulate systems, such as polymeric and lipid based nanoparticles, nanoemulsion, nanosuspension, complexation, solid dispersion and cosolvency have been widely used to circumvent above issues (Ud Din F et al., 2015a; Ud Din F et al., 2015b; Ud Din F et al., 2017; Rashid R et al., 2015a; Rashid R et al., 2015b; Ahmad MZ et al., 2016; Mustapha O et al., 2016). However, there is more focus on the vescicular systems, such as liposomes and niosomes as these systems are capable of encapsulating both lipophilic and hydrophilic drugs demonstrate sustained drug release, permeation enhancement and finally drug targeting minimizing toxic effect of drug (Abdelkader H et al., 2012). Liposomes are having membrane–mimetic structure because they are prepared from natural phospholipid lecithin. Liposomes are microscopic, generally ≤ 400 nm, vesicular structure enclosing as aqueous core surrounding by lipid bilayer of amphiphilic molecules. Apart from encapsulating both hydrophilic and lipophilic drugs, liposomes are having other advantages, such as enhance the drug permeation through membrane, avoid undesirable side effects, higher targeting capability (Mudshinge SR et al., 2011; Cevc G, 1997; Goyal P et al., 2005). However, liposomes suffer chemical and physical stability (Wang L et al., 2015).

Unlike liposomes, niosomes are stable and therefore have long storage time. They are made up of lipids, such as cholesterol as and nonionic surfactant, such as tween and span. Based on the size, niosomes are of three types: (i) small unilamellar vesicles (SUV) having size range from 10 to 100 nm, (ii) large unilamellar vesicles (LUV) of size in between 100 to 3000 nm and (iii) multi-lamellar vesicles (MLV) where more than one bilayer is present (Kaur IP et al., 2004). Niosomes are widely used as drug delivery system for the treatment of various diseases due to following advantages:

Advantages (Moghassemi S and Hadjizadeh A, 2014; Gandhi M et al., 2014)

1. Niosomes have long storage time when compared with liposomes as they are chemically stable.
2. Drug molecules of wide range of solubilities can be incorporated in niosomes, including hydrophilic, hydrophobic and amphiphilic.

3. Characteristics of niosomes can be controlled by altering composition, lamellarity size, surface charge and concentration of components.

4. Because of nonionic nature niosomes are highly compatible with biological systems and have low toxicity.

5. Niosomes can protect the drugs from harsh biological environment thereby improving their therapeutic performance.

6. They are capable of releasing drug in sustained manner.

7. They are nonimmunogenic and biodegradable.

8. They increase the oral bioavailability and enhance skin permeation of poorly soluble drugs.

9. Because of water-based suspension of niosomes, they offer high patient compliance.

10. They exclude phospholipids as a component, therefore special precautions and conditions are not essential.

11. Their large scale production is economical.

12. They can be employed as depot formulation, therby controlling the drug release.

Disadvantages

1. Like other drug carriers, niosomes are having disadvantages, such as:
Physical istability,

2. Possibility of aggregation and fusion,

3. Leakage of entrapped drug, and

4. Limiting the shelf-life of the dispersion due to hydrolysis of encapsulated drugs.

3.4.1   Preparation

Niosomal preparation is usually less tedious when compared with liposomes as the surfactant in niosomes resist against air oxidation. All the methods described below are consisting of a hydration of a lipid and surfactant mixture at elevated temperature (transition temperature of surfactant) and another optional step of size reduction to get colloidal dispersion.

3.4.1.1   Hand shaking method

This method involves mixing of surfactants and some additives, such as cholesterol in anorganic solvent in a round bottom flask followed by the removal of organic solvent employing rotary evaporator to form a thin film on the inside wall of the flask. The resulted completely dried film is then hydrated with aqueous solution of drug for almost a hour with mechanical shaking to form niosomal dispersion. Using this method, MLV can be prepared and this method is almost similar to thin film hydration (TFH) method (Moghassemi S and Hadjizadeh A, 2014).

3.4.1.2   Proniosome application

Proniosomes are resulted from the coating of water-soluble carrier, such as mannitol, sucrose stearates or maltodextrin with thin film of nonionic surfactants. Thereafter, depending on requirement these proniosomes are converted to niosomes by hydrating it at a temperature above transition temperature of surfactant. Therefore, these particles are dried particles encapsulating a water soluble drug carrier which improves its physical instability associated with niosomes, including fusion, aggregation and leakage thereby increase the drug entrapment efficiency. Furthermore, proniosomes can be further processed to obtain various unit dosage form, such as tablets, capsules etc. as it is obtained in the dry powder form (Marianecci C et al., 2014; Haghiralsadat F et al., 2018).

3.4.1.3   Heating method

In this method, vesicles components, such as surfactants and some additives, such as cholesterol are hydrated separately under nitrogen atmosphere in phosphate buffer saline (pH 7.4) for 1 hr at ambient temperature. Then, the above solution is heated at about 120°C on a hot-pate magnetic stirrer at a speed less than 1000 rpm in order to dissolve cholesterol. Thereafter, the temperature is reduced to transition temperature of the surfactant and other components including surfactants, drug and other excipients are then added to the buffer with stirring for 15 minutes in order to trap drug molecules in vesicles. The resulted niosomes obtained are kept at room temperature for 30 min and then at 4–5°C under nitrogen atmosphere prior to use (Mortazavi SM et al., 2007; Mozafari MR et al., 2005; Mozafari MR et al., 2007; Mozafari MR et al., 2002; Jahn A et al., 2004).

3.4.1.4   Transmembrane pH gradient method

This method involves mixing of surfactant and cholesterol in equal proportion in chloroform in a round-bottom flask and then evaporated under reduced pressure to generate a thin lipid film in the inside wall of flask. The resulted film is then hydrated with a solution of an acidic compound, usually citric acid, by vortex mixing and then subjected to freeze–thaw cycle (FT-cycle) followed by the addition of aqueous drug solution with vortexing. The pH of the solution is the adjusted to 7–7.2 (Mayer LD et al., 1995).

3.4.1.5   Thin-film hydration technique

TFH is widely used and simplest method to prepare niosomes. In this method, cholesterol and surfactants are dissolved in an organic solvent or a mixture of organic solvent in a round bottom flask followed by evaporation of organic solvent(s). This evaporation resulted in the formation of thin film inside wall of the flask. Then an aqueous drug solution or phosphate buffer saline (PBS) containing drug is slowly added to the flask above the transition temperature of the surfactant. The formed niosomes are removed and separated by means of techniques, such as centrifugation, dialysis or filtration. The niosomes can be lyophilized and packaged under aseptic condition prior to use (Shilpa et al., 2011; Bhaskaran S and Lakshmi PK, 2009; Arora R and Sharma A, 2010).

3.4.1.6   Freeze-drying

This method involves solubilization of cholesterol, surfactant, and any other additives in an organic solvent followed by the formation of thin film. Then, a buffer containing glucose or any other antifreeze material (cryoprotectant) for the hydration of thin film is added to above thin film. This resulted in the formation of MLV niosomes, which ware converted to SUV vesicles prior to freezedrying. After the freeze drying, these dried blank niosomes are placed in aqueous solution containing drug (Sankar V et al., 2010).

3.4.1.7   Reverse-phase evaporation technique

Unlike TFH, in reverse phase evaporation (REV) method involves both organic phase and aqueous phase are mixed without the formation of thin film. In the first step, organic phase is prepared by dissolving surfactants, cholesterol and other additives in an organic solvent. This step is followed by the preparation of aqueous phase containing drug and then both the phases are sonicated to form w/o emulsion. Finally, the organic phase is evaporated in a rotary evaporator under vacuum at 40-60 C, which resulted in the formation of LUV (Marwa A et al., 2013; Abdelkader H et al., 2011; Guinedi AS et al., 2005).

3.4.1.8   Freeze–thaw method

This method utilizes the empty niosomes prepared by thin film hydration method. The size of the niosomal suspension was reduced and then kept in liquid nitrogen (-196°C) for 5 min with a desired concentration of drug (Freeze). This is followed by transfer of niosomes into a water bath at surfactant transition temperature for 5 min (Thaw). These freeze and thaw steps are repeated 2 to 4 times which resulted in efficient entrapment of drug in the vesicles (Haghiralsadat F et al., 2018).

3.4.1.9   Dehydration rehydration method

In 1984, Kirby and Gregoria described dehydration and rehydration (DR) method. The vesicles prepared by TFH method is frozen in liquid nitrogen followed by freeze drying overnight. Thereafter, niosomal powders are hydrated with phosphate buffer saline (PBS) at pH 7.4 at 60°C (Hope MJ et al., 1986).

3.4.1.10   Ether injection

In this method, both aqueous phase and organic phases are prepared. The drug and surfactant are added to the phase in which it is soluble. For instance, drug is added to diethyl ether if it is hydrophobic and vice versa. The lipids and hydrophobic drugs are dissolved in diethyl ether and injected slowly into aqueous phase containing water soluble surfactant maintained at transition temperature of surfactant. The organic phase is then evaporated in a rotary evaporator or under vacuum (Haghiralsadat F et al., 2018; Marwa A et al., 2013). The resulted monolayer vesicles are niosomes with size ranging from 50 to 1000 nm (Uchegbu FI and Vyas SP, 1998; Verma S et al., 2010).

3.4.1.11   Sonication method

Sonication method involves the mixing of drug solution in buffer with cholesterol and surfactant mixture. The resulted mixture is then probe sonicated at the surfactant transition temperature or at 60°C with a sonicator for 3 min to obtain niosomes (Verma S et al., 2010).

3.4.1.12   Bubble method

Bubble method produces niosomes without the use of organic solvents. In the first step, surfactants and additives are mixed in phosphate buffer saline (pH 7.4) under nitrogen atmosphere at 70°C. After mixing for 15 s with high speed homogenizer, nitrogen bubbles are passed through it at 70°C, which resulted in the formation of niosomes (Moghassemi S and Hadjizadeh A, 2014; Verma S et al., 2010).

3.4.1.13   The Handjani–Vila method

In this method a homogenous lamellar phase is produced by shaking equivalent amount of nonionic lipids or lipid mixture with the aqueous solution of active ingredients. The resulted mixture is then homogenized at a controlled temperature employing agitation or ultracentrifugation (Alemayehu T et al., 2010).

3.4.1.14   The enzymatic method

This is another method to prepare multilamellar niosomes. It involves the use of enzyme, such as esterases to break eater links to produce intermediates, such as cholesterol and polyoxyethylene. These breakdown products in combination with dicetyl phosphate (or any other lipids) and surfactants, including polyoxyethylene cholesteryl sebacetate diacetate and polyoxyethylene steryl derivatives used to produce niosomes (Kumarn GP and Rajeshwarrao P, 2011).

3.4.1.15   Single pass technique

This technique involves both homogenization and high pressure extrusion in a single pass to produce niosomes. In this technique, lipid solution or suspension is passed through a porous device followed by through a nozzle. This technique has the capacity to produce niosomes within narrow size distribution range of 50-500 nm (Michael W et al., 2010).

3.4.1.16   Supercritical carbon dioxide fluid technique

A fluid is considered as supercritical when the pressure and temperature exceed its critical values (critical temperature and critical pressure. Beyond the critical temperature, it is impossible to liquefy a gas by increasing pressure. Out of the various gases used as supercritical fluid (SCF), CO_2 is widely used as it is having low critical values (31.3°C and 7.4 MPa) along with other favourable qualities, such as inertness, inexpensive, noninflammable etc. (Parhi R and Suresh P, 2013). Among various processes, rapid expansion of supercritical solution (RESS) is used to prepare vesicle systems. RESS process is based on the principle of supersaturation to form vesicle. It has two steps: dissolving all the solid substances in SC–CO2 and then resulted solution is depressurized through a heated nozzle air supersonic speed in to a low pressure chamber to form the particles (Parhi R and Suresh P, 2013). Manosroi first described about the preparation of niosomes using SCF (Manosroi A et al., 2008).

3.4.1.17   Microfluidization technique

Microfluidization method involves high velocity jets in micrometer scale. The lipid, surfactant and other additives in organic solvents is made to pass through a central channel and aqueous solution containing drug is introduced from two adjacent channels at a ultrahigh velocities. Both the phases are concentrated at the interaction point wherein inflow velocities of both the phases determine flow concentrations and led to interface formation. From the interaction chamber, the solution is made to pass through a cooling loop to dissipate the heat generated during the liquid movement (Verma S et al., 2010).

3.4.2   Application

Like other nanoparticulate systems, niosomes are versatile drug delivery system having various pharmaceutical applications. Some of the widely used applications are described below.

3.4.2.1   Oral delivery

Oral route is the most preferred route of drug administration because of it is most convenient, simple, provides the flexibility of accommodating varieties of formulations and most suitable to treat chronic diseases (Morishita M and Peppas NA, 2006; Peppas NA et al., 2004; Singh BN and Kim KH, 2000). Furthermore, it decreases the cost of treatment and also not involving pain during administration when compared with injection-based drug administration. However, the majority of existing and newly discovered drugs administered by oral route most often face bioavailability problem due to many reasons including low aqueous solubility, poor dissolution rate, existence of absorption window, inter and intrapatient variability, etc. Several approaches have been adopted for the improvement of above parameters, including mucoadhesive systems which will adhere to the vast mucous layer available across the gastrointestinal tract (GIT), micro- and nanoparticles (microspheres, polymeric and lipid based nanoparticles, vesicle systems) to increase surface area, complexation, solid dispersion, microencapsulation, etc. Among all, niosomes as drug delivery vehicles seems to be better alternatives because of the ease of preparation and can encapsulate either hydrophilic and lipophilic drugs apart from its nanosize which increases the surface area thereby improves drug absorption and subsequently bioavailability.

3.4.2.2   Pulmonary delivery

When compared with oral route, drug administration through pulmonary route is preferred for the prophylaxis and treatment of various local diseases, including asthma, hypertension, and other lung infections. This route can also be used to deliver drug molecules to exert systemic activity owning to its huge surface area covered with mucous, high vascularization, avoidance of first-pass metabolism and lung has very poor metabolic activity. However, the biggest challenge in pulmonary drug delivery is to reach the administered drug to the desired site (receptor) which is localized within the cytoplasm of bronchial epithelial cells. In these context particulate systems including niosomes is considered effective not only due to higher exposed surface area but also more importantly the requirements of particle size below 3 μm (Jyothi NV et al., 2010). Niosomes could amplify the therapeutic effect of loaded drug by improving mucous permeation along with sustained and targeted delivery (Terzano C et al., 2005).

3.4.2.3   Ocular delivery

In a recent report, it was estimated that nearly 256 million of world population have poor vision and almost 39 million people suffering from blindness (Rupenthal ID and O'Rourke M, 2016). Conventional dosage form, such as topical eyedrops is limited to their use in common diseases in anterior segments of eye including glaucoma, cataract and conjunctivitis. There are many factors, such as nasolacrymal drainage, tear turnover, poor permeation across the cornea and reflex blinking limits the desired drug concentration to reach target site to 5% of the administered dose (de la Fuente M et al., 2010). There are two strategies available in order to circumvent this issue, namely: (i) by increasing corneal residence time and (ii) employing penetration enhancers (Rupenthal ID and O'Rourke M, 2016). Out of the two strategies, the former looks more feasible as per as niosomes is concerned. This is because niosomes is capable of decreasing systemic drainage and improves the residence time, thereby increases ocular bioavailability of incorporated drug (Patel PB et al., 2010). In addition, bioadhesive-coated niosomal formulation can also improve the residence time due to mucoadhesive bond formation between bioadhesive polymer in the coating and mucin present on the eye surface (Dhiman S and Arora S, 2012).

3.4.2.4   Dermal and transdermal delivery

Administration of drug products topically via skin are categorized into dermal and transdermal drug delivery. In case of dermal drug delivery, the drug from the topical products delivered into the skin for the treatment of skin ailments. Thus, it improves the localization of drug and minimizes systemic absorption. Although the transdermal drug delivery utilizes skin as an alternative route for the delivery of drug into systemic circulation (Parhi R et al., 2015). The transdermal route of drug administration demonstrated many advantages over oral route including avoidance of presystemic metabolism in liver and GIT, no effect of variable pH and food–drug interaction in GIT, feasibility to release the drug in controlled or sustained manner for the drugs having short half-life, patient compliance. When compared with the parenteral route, transdermal route lacks pain during administration; possible infections after application thereby improve patient compliance (Joshi A and Raje J, 2002; Cummings EA et al., 1996). Despite all, there are only few transdermal products in the market which is due to the impervious nature of stratum corneum (SC), uppermost layer of skin. There are many penetration enhancement technique, such as iontophoresis, ultrasound, magnetophoresis, electroporation laser technique and more recently microneedle technique. However, all these techniques are having their own drawbacks. Since last decade or so vesicle drug deliver across the skin has been the topic of intense research. Niosomes can fuse with the SC and may contribute to the permeation enhancement of drug incorporated in it. Niosomes as a formulation cannot be applied on to the skin surface, instead a carrier is essential. Ointments, creams and gels could be used as carrier for niosomes so that it can be applied on the skin surface which will adhere to it for considerable period of time.

3.4.2.5   Parenteral delivery

Drug delivery using parenteral route has certain advantages, such as rapid onset of action, highest bioavailability (100% in case of intravenous route), favourable in conditions when oral route is not convenient, such as vomiting, difficult in swallowing and unconscious situations, avoidance of first-pass metabolism and harsh environment of GIT (Khatoon M et al., 2017). However, frequent injections are required to sustain the therapeutic action leading to low patient compliance is major challenge. Niosomes as a drug delivery system capable of providing not only sustained drug release, but also to target various affected sites in the body. Sustained release can be achieved by injecting niosomal formulations through subcutaneous and intramuscular route (Marianecci C et al., 2014).

3.4.2.6   Gene delivery

Gene therapy is considered as the therapy of the 21st century as it has the potential to treat almost all diseases. It involves the administration of genetic material, such as DNA, oligonucleotides, ribozymes, DNAzymes and small RNAs into or body cells. However, during the delivery of genetic materials, the issues, such as stability, cellular uptake, intracellular trafficking and alteration of biodistribution limit its efficiency (Hood E et al., 2007; Hong M et al., 2009). Therefore, the use of nonionic niosomes as a carrier for genetic material as the surfactant in the niosomes enhances the stability of genetic material and also plays an important role in the biological integration with genetic material, such as DNA resulting in better gene expression (Marianecci C et al., 2014).

3.4.2.7   Diagnostics/theranostics

Nanocarriers, such as niosomes can be used successfully to improve the delivery of imaging agents for tumours, which in term helps in diagnosis at early stages. The principle of PEGylation, i.e. PEG conjugated to the surface of niosomes encapsulated with paramagnetic agent assessed with nuclear magnetic field (NMR) imaging in a carcinoma was found to be successful

(Luciani A et al., 2004). Various applications of niosomes in different routes along with the composition are listed in Table 3.4.

Table 3.4: Different applications of niosomes (Marianecci C et al., 2014)

Applications	Drug	Composition	Results
Oral delivery	Ganciclovir	Span 40/Span 60 and Cholesterol	Fivefold increase in bioavailability of orally administered optimized niosomes (Span60:Cholesterol, 3:2) compared with tablet in rat model
	Paclitaxel	Tween/Span/Brijs and cholesterol and Dicetyl phosphate (DCP)	Best drug protection from GI enzymes with Span 40 based niosomes and fast drug release from tween based niosomes
	Oxcarbazepine	Spans and Cholesterol	Higher elimination half-life and area under the curve (AUC) compared with free drug
	Lornoxicam	Spans and Cholesterol	Prolonged and consistent release of drug
	Valsartan	Span60, Cholesterol, maltodextrin: proniosomes	Drug dissolution is increased and improved drug absorption across GI membrane
Pulmonary delivery	Beclomethasone dipropionate	Tween20 and Cholesterol	Increased permeation rate across model mucosal barrier and anti-inflammatory activity in human lung fibroblast cells (HLFs). Higher degree of tolerability on HLFs
	Beclomethasone dipropionate	Span 60, Cholesterol and Sucrose: Proniosomes	Increased fine particle fraction (FPF) by all devices with higher drug output.
	Ciprofloxacin HCl	Span 60/ Tween 60 and Cholesterol	High degree of *in vitro* safety and efficacy of niosomes loaded with ciprofloxacin
	Amphoteracin B	Tween 80 and cholesterol with hydroxypropyl-γ-cyclodextrin	Reduced fungal lung burden in invasive pulmonary aspergillosis in rat model significantly.
Ocular delivery	Gentamycin	Tween 60/Tween 80/Brij 35, and Cholesterol and DCP	Higher the chain length of alkyl, the lower the drug release, DCP delays the drug release and stabilizes the niosomal membrane
	Timolol maleate	Span 60 and Cholesterol (1:1) coated with chitosan or carbopol	Significantly reduced intraocular pressure (IOP)
	Naltrexone HCl	Span60 and DCP/ Sodium Cholate	Improved protection of drug from photo-oxidation and enhanced corneal uptake compared with aqueous solutions of the drug
	Cyclopentolate	Tween 20 and Cholesterol	Enhanced in vitro transcorneal permeation and ocular bioavailability

Contd...

Contd...

Applications	Drug	Composition	Results
Dermal and transdermal delivery	Tretinoin	Span 40/60/ Brij30/Triton CG 110, cholesterol and DCP	Higher drug permeation and local drug concentration
	Enoxacin	Span40/60, cholesterol and/ or DCP	Higher local drug concentration
	Meloxicam	Span 60 and Cholesterol based niosomes in hydrogel	Niosomal gel significantly increase in the inhibition of oedema in animals
	Lopinavir	Span 40 and cholesterol based niosomal gel	Higher bioavailability and higher deposition with better safety compared with ethosomal gel
Parenteral delivery	Cisplatin	Span 40 and Cholesterol based freeze dried niosomes	Lower the mortality compared with free drug due to significant inhibition of tumour growth and metastasis in lymph nodes and liver
	Nystatin	Span 60/40, Cholesterol, stearylamine and/ or DCP	Enhanced antifungal activity with reduced drug toxicity
	Paclitaxel	Nanohybrid system-Solutol HS 15, pluronic F68 and cremophor EL with inserted liposomes loaded with drug	Empty nanohybrid interfere P-gp expression on the membrane of drug resistant cells and decrease ATP level. Higher cytotoxicity compared with liposomes
Gene delivery	Small interfering RNA	Span 80 and cationic lipid	Silence expression of both green fluorescent protein and aromatase genes in breast cancer cell lines
	Plasmids	Tween 61, cholesterol, dimethyl dioctadecyl ammonium bromide (1:1:0.5)	Enhancement of tyrosinase gene expression and melanin production in human human melanoma cells and tyrosine producing mouse melanoma cells (B12F10)
Diagnostics	With ultrasound	Tween 61, Cholesterol and DCP	Ultrasound at specific frequencies can reversibly permeabilize the lipophilic membrane of niosomes to allow the controlled release of a compound without destroying the structure
	Lipophilic and hydrophilic magnetic nanoparticles	Tween 20.Span 20 and Cholesterol	Theranostic properties

3.5   Aquasomes

Aquasomes are spherical solid nanoparticulate drug carriers with particles size ranging from 60 to 300 nm. Aquasomes are three layered self-assembly structures, composed of solid phase nanocrystalline particulate core with surface coated with glassy carbohydrate molecules, which is then adsorbed with therapeutic molecules. These three layered structures are assembled by ionic and noncovalent bonds. There are three main types of core materials, such as carbon ceramics, tin oxides and calcium phosphate dehydrate (brushite). The core structure is stabilized by surface modification using carbohydrates, such as cellobiose. The pharmacologically active drug molecules including small drug molecule, large drug molecules (e.g. proteins and peptides based hormones), enzymes, antigens and genes can not only be delivered in desired concentration but also delivered at the proper site of action (drug targeting). These actives are adsorbed or copolymerized or diffused into carbohydrate layer with or without modification and are completely protected and preserved against denaturation effects of external temperature and pH as there are no porosity and swelling changes in accordance with the change in temperature and pH (Godin B and Touitou E, 2003; Vyas SP and Khar RK, 2002; Luo D et al., 2004; Umashankar MS et al., 2010).

3.5.1   Principle of Self-Assembly

In any product, if the constituent parts assume spontaneously prescribed structural orientation in two or three dimensional space is called as self-assembly. The self-assembly of molecules in the aqueous environment is usually governed by three physicochemical processes, including the interaction of charged groups, dehydration effect and structural stability.

3.5.1.1   Interaction between charged groups

Surface of most of the synthetic and biological molecules are charged due to intrinsic chemical groups or adsorbed ions from the surrounding. The interaction of charged groups, such as amine, sulphate, carboxyl, phosphate groups enhance the formation of self-assembly subunits. Thereafter, long range interactions of these subunits and their gradual attraction are essential for the first phase of self-assembly.

3.5.1.2   Hydrogen bonding and dehydration effects

Molecules forming hydrogen bonds are hydrophilic which confers a significant degree of organization to surrounding water molecules. However, hydrophobic molecules are unable to form hydrogen bond. But, their water repelling tendency facilitates to organize the units to surrounding environment. The formed organized water reduces the overall level of disorder of the surrounding medium. As it is well known that the organized water is thermodynamically unfavourable, the selected molecule loose water (dehydrate) and get assembled (Kossovsky N et al., 1994).

3.5.1.3   *Structural stability*

Dipole moment is exhibited by molecules which carry less charge compared molecules bearing formal charged group. The force generated from dipole moment is called as van der waals forces. In biological environment structural stability of protein determined by interaction between charged group and H-bonding largely external to molecule and van der waals forces largely internal to molecule. This van der waals forces most often experienced by hydrophobic regions of a molecules and this force shielded the molecule from water. This resulted in maintenance of molecular shape or conformation during self-assembly. Therefore, van der waals forces are mainly responsible for softness and hardness of any molecules. Interaction of this force among the hydrophobic side chain promotes stability of compact helical structures which are thermodynamically unstable for expanded random coils. Therefore, it helps in the maintenance of internal secondary structures, such as helices that allows sufficient softness and allows maintainance of conformation during self-assembly (Jain NK, 2001).

3.5.2   Preparation

Using the principle of self-assembly, the aquasomes are prepared in following three steps: (1) preparation of core, (2) coating of core, and (3) loading of drug molecule (Kossovsky N et al., 1990; Kossovsky N and Millett D, 1991; Kossovsky N et al., 1994). The preparation steps are depicted in Figure 3.5.

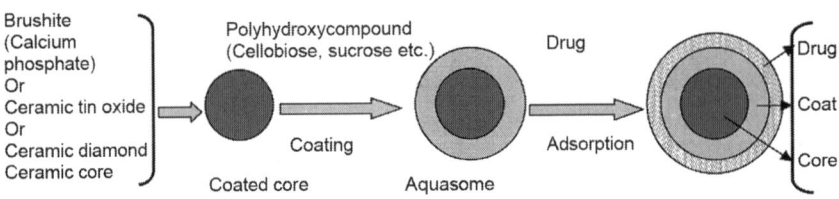

Figure 3.5: Preparation steps of aquasomes

3.5.2.1   Preparation of core

This step involves the fabrication of ceramic core. The process of ceramic core development is depends on the selection of core material. Ceramic materials are the choice of material due to their higher degree of crystalline order that ensures a minimum effect of surface modification on the atoms present in the inner layers, thereby the bulk properties of the material is preserved. Furthermore, higher degree of order in the ceramics provides a higher degree of surface energy which felicitates the bonding of polyhydroxy oligomeric film to it. There are three types of ceramic materials used to prepare core including tin oxide, brushite and diamond. The core can be fabricated using various methods, such as colloidal precipitation, sonication, inverted magnetron sputtering and plasma condensation methods. Following core particles are crystalline with clean surface and having measured size between 50 and 150 nm. The clean surface renders it more reactive and used for the next step of coating (Jain SS et al., 2012).

3.5.2.1.1   Self-assembled nanocrystalline brushite (calcium phosphate dehydrate)

It can be synthesized by colloidal precipitation and sonication by reacting solution of disodium hydrogen phosphate and calcium chloride (equation). The resulted precipitated cores are centrifuged followed by washing with enough amount of distilled water in order to remove sodium chloride generated during the reaction. Subsequently, the above precipitated core is resuspended in distilled water and passed through a fine membrane filter to separate desired size particles (Jain SS et al., 2012).

$$2Na_2HPO_4 + 3CaCl_2 + H_2O = Ca_3(PO_4)_2 + 4NaCl + 2H_2 + Cl_2 + (O)$$

3.5.2.1.2   Synthesis of nanocrystalline tin oxide core ceramic

It can be prepared by employing direct current reactive magnetron sputtering. In this method, a target of high purity tin, diameter of 3 inches is sputtered in a gas mixture of argon and oxygen present at high pressure. This lead to the generation of ultrafine particles in the gas phase which is then collected on copper tube which is cooled to 77°K with flowing nitrogen.

3.5.2.1.3   Nanocrystalline carbon ceramic, diamond particles

Diamond particles can also be used for the core synthesis following ultracleansing and sonication.

3.5.2.2   Coating of the core

Once the ceramic core of desired size is prepared, these are then subjected to coating with suitable material, such as carbohydrate (polyhydroxyl oligomer).

This is carried out in two steps: in the first step the carbohydrate is added to aqueous dispersion of core and mixed thoroughly under sonication, and then the resulted dispersion is subjected to lyophilization to improve irreversible adsorption of carbohydrate molecules onto the ceramic core. The remaining unadsorbed carbohydrate molecules are separated by centrifugation. The carbohydrates, such as cellobiose, sucrose and trehalose are commonly used to coat ceramic core (Jain SS et al., 2012).

3.5.2.2.1   Cellobiose

It is a disaccharide comprises two glucose units with beta (1→4) glycosidic bond. It is obtained from cellulose by enzymatic or acidic hydrolysis. It has eight free alcohol ([sbond]OH) groups, one acetal linkage and one haemiacetal linkage which is helping in bonding with central core and drug by hydrogen bonding. It is soluble in water and almost insoluble in alcohol and ether.

3.5.2.2.2   Trehalose

It is a sugar formed by union of two alpha-d glucopyranose molecule. It is also called as tremalose or mycose and obtained from some bacteria, fungi, plants and invertebrate animals. For insects, it is the source of energy and to survive freezing and inadequate water condition. It gets hydrolyzed by trehalose or mineral acid into glucopyranose units. It is soluble in water and insoluble in ether.

3.5.2.2.3   Sucrose

Sucrose is a disaccharide comprises one glucose unit and one fructose unit through an ether bond between C1 of alpha-d-glucopyranose molecule and C2 of beta-d-fructopyranose molecule. It can be hydrolyzed by dilute mineral acid on heating and invertase enzyme.

3.5.2.3   Loading of drug molecules

This is the final step in the preparation of aquasomes where the coated particles are loaded with drug by adsorption. It involves the preparation of drug solution of known concentration in a suitable pH buffer and the coated particles are dispersed in it. For proper adsorption of drug, the dispersion is either kept overnight at low temperature or lyophilized after sometime to get loaded coated particles (aquasomes) (Jain SS et al., 2012).

3.5.3   Applications

3.5.3.1   Delivery of poorly soluble drugs

Like other vesicular systems, aquasomes are widely used in the delivery of both hydrophilic as well as poorly soluble drugs. Piroxicam, a poorly soluble drug, was loaded on aquasomes with ceramic core and trehalose coating. These aquasomes exhibited controlled release of piroxicam, whereas nanoparticles without trehalose showed 90% of drug release in 1 h (Vengala P et al., 2013a; Banerjee S and Sen KK, 2018). Another low soluble drug indomethacin was adsorbed on lactose film coated over the calcium phosphate core (Oviedo RI et al., 2007). Similarly, calcium phosphate core coated with lactose used to deliver hydrophobic drug pimozide. The release of pimozide from the aquasomes exhibited first order kinetics (Vengala P et al., 2013b). Aquasomes with organ targeting potential was successfully developed for etoposide. *In vivo* study showed that the maximum amount of injected dose of drug was in liver followed by spleen (Nanjwade BK et al., 2013).

3.5.3.2   Insulin delivery

Aquasomes can provide an attractive platform not only to deliver insulin for prolonged period of time, but also to maintain the structural integrity of insulin. The prolonged activity is attributed to the slow release of insulin from the aquasomes and structural integrity is due to prevention of denaturation and dehydration of insulin (Paul W and Sharma CP, 2001). Insulin loaded aquasomes were developed using calcium phosphate as ceramic core, disaccharides, such as trehalose, cellobiose and pyridoxal-5-phosphate as coating material, and then the resulted aquasomes were adsorbed with insulin. Among all the coating material, pyridoxal-5-phosphate was found to show more activity in reducing blood glucose levels. This was probably due to high degree of molecular preservation (Cherian AK et al., 2003). An optimum controlled release of insulin was observed in case of aquasomes prepared from porous hydroxyapatite nanoparticles as core and alginate as coating material (Paul W and Sharma CP, 2001).

3.5.3.3   Enzyme delivery

Aquasomes are also used to deliver enzymes, such as acid-labile enzyme serratiopeptidase and DNAase by protecting their structural integrity, thereby providing their better therapeutic effect. In case of serratiopetidase, the nanocore was coated with chitosan and then enzyme was adsorbed on to its surface. The resulted enzyme-loaded aquasomes was dispersed in alginate gel to protect the enzyme from direct exposure to acidic medium. The *in vitro* drug release study exhibited Higuchi model in acid medium up to 6 hr, whereas first-order release of the enzyme was observed in alkaline medium for up to 6 hr (Rawat M et al., 2008). The fluctuation of molecular conformation can be prevented by loading DNAase on aquasome. In one study, the DNAase enzyme was loaded on polyhydroxyl oligomeric films present over calcium phosphate nanocore. The resulted aquasomes loaded with DNAase showed higher biological activity against cystic fibrosis (Vays SP and Khar RK, 2004).

3.5.3.4   As oxygen carrier

Aquasomes has the ability to deliver large and complex labile molecules, such as haemoglobin. This is possible due to the adsorption of haemoglobin on to the surface of aquasomes. In one such case, the core of hydroxyapatite modified with carboxylic acid terminated half-generation poly(amidoamine) dendrimers was coated with trehalose followed by heamoglobon adsorption (Khopade AJ et al., 2002). Above aquasomes exhibited good loading capacity ($\sim$ 13.7 mg of haemoglobin per gram of core) and oxygen retention. It was found to be stable for 30 days without any sign of haemolysis. In another study aquasomes with hydroxyapatite as core and cellobiose, trehalose, maltose, and sucrose as coating material were developed and then adsorbed with heamoglobin. The heamoglobin adsorbed on these aquasomes did not show any type of instability for 30 days and oxygen carrying capacity and delivery ability (nonlinear manner) of these aquasome were found to be similar to that of fresh heamoglobin (Patil S et al., 2004). Therefore, the issues, such as incompatibility and storage can be solved using heamoglobin loaded aquasomes.

3.5.3.5   Antigen delivery

The adjuvants are used along with antigens to boost their immunity. However, adjuvants are having tendency to alter the conformation of the antigen through surface adsorption. Therefore, an appropriate carrier, such as aquasomes which can deliver the antigen to the proper site by preventing it from conformational alteration is of great importance. In one of the study, diamond nanocores were coated with cellobiose and then antigen (muscle adhesive protein) is adsorbed on to it. Here, cellobiose inhibits denaturation of adsorbed antigen, thus acted as natural conformational stabilizer. In addition, it provides high degree of surface exposure to antigen protein. The resulted antigen loaded aquasomes exhibited strong and specific immune response when compared with antigen without aquasome carrier. This may be due to the enhancement of availability and *in vivo* activity of antigen (Kossovsky N et al., 1995a). In another investigation the immobilized antigen (nonnuclear material from human immunodeficiency virus-1, HIV-1) on carbon and calcium phosphate nanocore showed cellular and humoural immune responses similar to that of elicited by live HIV (Kossovsky N et al., 1995b). Bovin serum albumin (BSA) as antigen was adsorbed on to the surface of aquasomes made up of hydroxyapatite as core and cellobiose and trehalose as coating materials. The immunological activity of the resulted formulation exhibited better response when compared with plain BSA (Vyas SP et al., 2008).

3.5.3.6   Vaccine delivery

Aquasomes can be used to deliver vaccine, such as hepatitis B, Epstein-Barr virus (EPV) and HIV for immunoprotection. A nanodecoy system was used to delivery of Hepatitis B vaccine delivery. It involves the preparation of a self-assembled hydroxyapatite core followed by coating of cellobiose. The hepatitis B surface antigen was adsorbed over the coated core. These nanodecoy preparations exhibited increased immune system (Goyal AK et al., 2006).

Aquasomes with HIV vaccine is developed by using surface modified carbon and calcium phosphate as core, cellobiose as coating material and then mixed with emulsified viral protein. These aquasomes could develop both cellular and humoural immune response very much similar to that produced by whole or live HIV virus (Kossovsky N et al., 1995b).

In case of EPV adsorbed aquasomes, first the glycoprotein-350 is isolated from the EPV virus and then adsorbed onto the cellobiose coating present above tin oxide core. This self-assembled viral nanodecoy showed both immunological and physiological similarities with that of live EPV. Furthermore, this could elicit four times greater response when compared with pure viral envelop.

3.5.3.7   Gene delivery

Aquasomes seems to be an attractive delivery system for genetic material as it protect and maintain the structural integrity of the gene segments. In addition, aquasome vehicle can avoid the risk of irrelevant gene integration (Kossovsky N et al., 1994). A five layered aquasome has been proposed for gene therapy. The five layers are: ceramic core of nanosize, a polyhydroxyl oligomeric film coating, layer of gene segment with noncovalent bonding, carbohydrate film and finally viral membrane protein with specific targeting potential (Jain SS et al., 2012).

3.6   Phytosomes or Herbosomes

The terms "phyto" or "herbo" mean plant, while "some" means cell-like. Phytosomes or herbosomes are better alternative carriers to liposome in delivering pharmaceutical ingredients, especially phytoconstituents. This is because phytosomes avoid the use of cholesterol that creates certain problems like stability by oxidizing itself when used in the treatment of artherosclerosis

(Samuni U et al., 2000; Goyal K et al., 2005). These are generated from the complexation of herbal drug with phospholipids, such as soy phospholipids. This complexation is based on the formation of hydrogen bonds between polar head of phospholipids and polar group of herbal drugs. This bonding is a very important property of phytosomes which provides them higher physical stability when compared with liposomes. These are first tried for their antiaeging and cosmetic applications. Presently, they are being used for delivery of pharmaceuticals for the treatment of liver diseases, inflammatory, cardiovascular problems and cancer (Freaga MS et al., 2018). The important advantages of phytosomes when compared with other pharmaceutical formulations are:

Advantages

1. Phytosomes enhance the absorption of hydrophilic botanical extract across oral as well as topical route due to their complexation with phospholipids, thereby exhibiting superior bioavailability and therapeutic benefits (Jain N et al., 2010).

2. They improve the absorption of phytoconstituents leading to the reduction in dose requirement (Karimi N et al., 2015).

3. Phytosomes felicitate the absorption of nonlipophilic botanical extract across intestinal lumen (Valenzuela A et al., 1989).

4. As all the components of phytosomes have been approved, therefore the phytosomes are safe for pharmaceutical and cosmetic use.

5. It adds nutritional benefits of phospholipids.

6. The main excipient of phytosomes is phosphatidylcholine, which not only acts as carrier for phytoconstituents but also acts as hepatoprotective, thus providing synergistc effect when hepatoprotective ingredients are incorporated into it (Saraf S and Kaur CD, 2010).

7. When compared with liposomes, phytosomes are more stable because of chemical bonds formed between phsphotidylechloine and phytoconstituents (Bansal S et al., 2012).

8. Drug entrapment in phytosomes has not been a problem as the drug itself forms vesicles after complexation with lipid that is biodegradable.

9. Toxicological profiles of the phytosomes components are well documented in the scientific literature, thus the technology has low risk of drug development (Battacharya S, 2009).

10. Relatively simple manufacturing process without complicated technical investment.

11. Being an essential component of phytosomes, phosphatidylcholine not only nourish the skin, but also felicitate the permeation pf drug in to skin and across skin in case of dermal and transdermal drug delivery (Amin T and Bhat S, 2012).

12. They are capable of delivering large and diverse group of drugs including protein and peptides.

3.6.1   Preparation

Solvent evaporation technique is widely used in the preparation of phytosomes. There are three components in the preparation of phytosomes including phospholipids, such as phosphatidylcholine, phosphatidylethanolamine and phosphatidylserine, phytoconstituents (i.e. flavonoids and terpenoids), and organic solvents, such as dioxane, acetone, methylene chloride or ethyl acetate. Preparation involves the reaction between natural or synthetic phospholipids and herbal extract in particular ratio. In the complex formation the ratio between these two components range from 0.5 to 2 moles with most preferable ratio of phospholipids to flavonoids is 1:1 (Jose MM and Bombardelli E, 1987). After the formation of complex, the solvent is evaporated under vacuum or lyophilization or spray drying. Precipitation method is also being used to prepare phytosomes. In this method, the formed complex is precipitated with nonsolvent, such as aliphatic hydrocarbons and then both the solvents are separated or evaporated to get thin layer (Singh D et al., 2012). This is followed by hydration and sonication led to the formation of phytosomes as depicted in Figure 3.6.

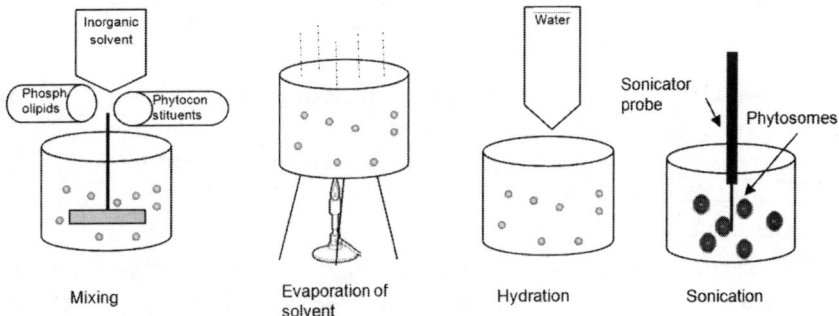

Figure 3.6: Schematic representation of steps involved in the preparation of phytosomes

3.6.2   Application

There are numerous plant drugs that are encapsulated into phytosomes. Among them, the phytoconstituents obtained from milk thistle plant (*S. marianum*, Family Steraceae) are studied intensively. The fruit contains three flavonoids with hepatoprotective effects, including silybin, silydianin and silychristin. Silybin, a flavonolignan, is the most among three and protects the liver by restoring glutathione in the parenchymal cells (Hiking H et al., 1984). Silymarin is used to treat various liver diseases, such as hepatitis, cirrhosis and inflammation of the bile duct. Apart from that, it also protects liver from toxic materials (Valenzuela A et al., 1989). Many researches had prepared phytosomes of silymarin and silybin. In one such study, silymarin phytosomes were prepared by complexing silymarin with phospholipids and the resulted phytosomes found to show better hepatoprotective activity when compared with silymarin alone. In addition, it can provide protection against the toxic effects of aflatoxin B1 when tested in broiler chicks (Ravarotto L, 2004). In another study, silymarin phytosome exhibited higher fetoprotectant activity when compared with uncomplexed silymarin (Busby A et al., 2002). Oral administration of silymarin phytosome showed has protective action on fetus from maternally ingested ethanol (La Grange L et al., 1999). The bioavailability of phytosome developed from the silybin and phospholipid complexation was increased notably after oral administration. This was attributed to the remarkable enhancement lipophilic property of phytosome (Yanyu X et al., 2006). A novel phytosome was developed by complexing hesperetin with hydrogenated phosphatidyl choline. This phytosomes exhibited higher antioxidant activity with sustained release property for over 24 h in carbon tetrachloride treated rats (Mukherjee K et al., 2008).

Phytosomes based on quercetin and phospholid complex exhibited better therapeutic efficacy in liver injury of rat induced by CCl4 (Maiti K et al., 2005).

Phytosomes are also prepared with grape seed extract (*Vitis vinifera*) containing proanthocyanidins (a flavonoid) and phopspholipids. Above phytosomes increased the antioxidant capacity and improved physiological antioxidant defenses of plasma, protective action against atherosclerosis and ischaemia, etc (Schwitters B and Masquelier J, 1993).

The extract of green tea leaves (*Thea sinensis*) contains epigallocatechin and its derivatives as polyphenols having antioxidant property. Apart from antioxidant activity, it also shows anticarciogenic, hypocholesterolaemia, antibacterial, anti-atherosclerotic and antimutagenic activities. However, tea polyphenol extract demonstrated very poor bioavailability. However, the formation of phytosomes by complexing the polyphenol with phospholipids improves their bioavailability. In one investigation, the phytosomal preparations were able to maintain plasma drug concentration of total polyphenol more than 2-fold when compared with nonphytosomal counterpart. Furthermore, the antioxidant property of phytosomal polyphenol was found to be double when compared with extract alone when measured as total radical-trapping antioxidant parameter.

Curcumin, a flavonoid, obtained from plant *Curcuma longa* exhibited higher antioxidant property in phytosomal form when compared with pure curcumin in all dose level tested (Maiti K et al., 2007). A lists of commercially available phytosomes with their trade name and uses are oresented in the Table 3.5.

Table 3.5: Various marketed phytosomeswith their phytoconstituents, source and biological activity (Karimi N et al., 2015)

Phytoconstituents	Source	Trade name	Biological activity
Anthocyanosides	*Vaccinium myrtillus*	Bilberry Phytosome	Antioxidant and improvement of capillary tone
Gum resins	*Banksia serrata*	CasperomeTM	Higher systemic availability and improving tissue distribution of boswellic acids
Terpenes	*Centella asiatica*	Centella phytosomes	Vein and skin disorder, brain tonic.
Polyphenol	*Curcuma Longa*	Curcumin phytosomes	Improving the bioavailability of curcuminoids and cancer chemo-preventive agent
Curbilene	Cucurbita pepo seeds	Curbilene phytosomes	Skin care, Matting agent.
Echinacosides	*Echinacea angustifolia*	Echinacea phytosomes	Neutraceuticals, Immunomodulatory.
	Echinacea purpurea		Immunomodulatory
Flavonoids	*Ginkgo biloba*	Ginkgo select phytosomes	Anti-aeging, Protection of vascular lining of brain
Ginsenosides	*Panax ginseng*	Ginseng phytosomes	Nutraceuticals, Immunomodulator
Procyanidins	*Vitis vinifera*	Grape seed (Leucoselect) phytosomes	Antioxidant, Anticancer, Neutraceuticals
Procyanidins A2	Chestnut bark	PA2 phytosomes	Antiwrinkles, UV protection
Flavonoids	Crataegus species	Hawthorn phytosomes	Cardioprotective, Antihypertensive
Triterpenes	*Melilotus officinalis*	Melilotus (Lymphaselect) phytosome	Hypotensive, Insomnia

Contd...

Contd...

Phytoconstituents	Source	Trade name	Biological activity
Polyphenols, Anticinoside	*Vaccinium myrtillus*	Mirtoselect phytosomes	Antioxidant
Fruit and leave extracts	Olive	OleaselectTM	Higher availability compared with crude extract
Fatty acids, sterols and alcohols	*Serenoa repens*	Palmetto phytosomes	Antioxidant, Benign prostate hyperplasia
Steroids, Saponins	*Ruscus aculeatus*	Ruscogenin Phytosomes	Improve skin circulation, Anti-inflammatory.
Sericosides	*Terminalia sericea*	Sericoside phytosomes	Antiwrinkles, Soothing
Silybin	*Silybum marianum*	Silybin phytosomes	Antioxidant, Phytoprotective
Visnadine	*Ammi visnaga*	Visnadine phytosomes	Improving blood circulation
Ximilene and ximen oils	*Santalum album*	Ximilene and ximen oils phytosomes	Micro-circulation improver, Skin smoother
Zanthalene	*Zanthoxylum bungeanum*	Zanthalene phytosomes	Anti-irritant, anti-itching, Soothing

3.7   Electrosomes

These are novel surface-display system consisting of two electrodes, such as anode and cathode. This is based on the specific interaction between the cellulosomal scaffoldin protein and a cascade of redox enzymes which allows multiple electrol release. In case of performance of biofuel, electrosome is formed of two components, such as hybrid anode and hybrid cathode. Further, the anode is composed of dockerin-containing enzymes adhere specifically to cohesion sites of scaffoldin to perform ethanol oxidation cascade, whereas hybrid cathode consisting of oxygen-reducing enzyme in dockerin that is attached in multiple copies to the cohesion-bearing scaffoldin. Electrosome was formed in both the compartments having cellulosome scaffoldin. The anode is having enzymes, such as alcohol dehydrogenase and formaldehyde dehydrogenase where ethanol oxidation cascade is formed. At cathode, copper oxidase enzyme expressed by *E. coli* was employed. This electrosome with new hybrid cell compartments displayed enhanced performance over traditional biofuel cells (Szczupak A et al., 2017).

3.8   Conclusions

The current chapter presented various particulate systems, such as microspheres, niosomes, aquasomes, phytosomes, and electrosomes with their types, methods of preparation, evaluation and applications. In addition, the concept, types, method of preparation and application of mAbs are also discussed. The evolution and popularity of particulate systems has been focused on three aspects, such as: need for less toxic drug carriers, simplification of preparation technique for economical scale-up using advanced technology and optimization by using quality by design to improve yield and better EE. Based on the above requirements, there are many particulate systems developed and optimized for their suitability to encapsulate entire gamut of biological actives and to deliver at the target site with desired concentration.

Not only that, particulate systems are being employed as diagnostic agents for *in vivo* imaging of tumour and other diseases. In the context of diagnosis, there has been major use of mAbs because of rapid advancement of genetic engineering. One study shows that the global value of the antibody market is $20 billion per year. There are many particulate drug carrier already approved by FDA and several others are presently in development and clinical stage. The use of technology in the development of particulate systems would yield in a concomitant improvement in the quality, affordability, efficacy and safety profile of dosage forms.

3.9   References

Abciximab-ucm107731.pdf.   https://www.   Accessdata.fda.   gov/drugsatfda_docs/ label/1997/abcicen110597-lab. pdf (Accessed Jan 09, 2018).

Abdelkader H, Ismail S, A. Kamal A, Alany R G (2011), 'Design and evaluation of controlledrelease niosomes and discomes for naltrexone hydrochloride ocular delivery',J Pharm Sci, 100, 1833–1846.

Abdelkader H, Wu Z, Al-Kassas R, Alany R G (2012), 'Niosomes and discomes for ocular delivery of naltrexone hydrochloride: morphological, rheological, spreading properties and photo-protective effects', Int J Pharm, 433, 142–8.

Acharya U H, Jeter J M (2013), 'Use of ipilimumab in the treatment of melanoma', Clin Pharmacol, 5(Suppl. 1), 21–7.

Ahagon A, Gent A (1975), 'Effect of interfacial bonding on the strength of adhesion', J Polym Sci Polym Phys Ed, 13, 1285–1300.

Ahmad M Z, Akhter S, Mohsin N, Abdel-Wahab BA, Ahmad J, Warsi MH, Rahman M, Mallick N, Ahmad F J (2014), 'Transformation of curcumin from food additive to multifunctional medicine: nanotechnology bridging the gap', CDDT, 11, 197–213.

Ahmad M Z, Alkahtani S A, Akhter S, Ahmad F J, Ahmad J, Akhtar M S, Mohsin N, Abdel-Wahab B A (2016), 'Progress in nanotechnology-based drug carrier in designing of curcumin nanomedicines for cancer therapy: current state-of-the-art', J Drug Targeting, 24, 273–93.

Alemayehu T, Nisha M J, Palani S, Anish Z, Zelalem A (2010), 'Niosomes in targeted drug delivery: some recent advances', Int J Pharm Sci Res, 1, 1–8.

Alinari L, Lapalombella R, Andritsos L, Baiocchi R A, Lin T S, Byrd J C (2007), 'Alemtuzumab (Campath-1H) in the treatment of chronic lymphocytic leukemia', Oncogene, 26, 3644–53.

Amin T, Bhat S (2012), 'A Review on phytosome technology as a novel approach to improve the bioavailability of neutraceuticals', Int J Adv Res Technol, 1, 1-15.

Arora R, Sharma A (2010), 'Release studies of ketoprofen niosome formulation', J Chem Pharm Res, 2, 79–82.

Arora S, Ali J, Ahuja A, Khar R K, Baboota S (2005), 'Floating Drug Delivery Systems: A Review', AAPS PharmSciTech, 6, E372-390.

Arshady R (1989), 'Microspheres and microcapsules: A survey of manufacturing techniques. Part 1: Suspension cross-linking', Polym Eng Sci, 29, 1746-58.

Banerjee S, Sen K K (2018), 'Aquasomes: A novel nanoparticulate drug carrier', J Drug Deliv Sci Technol, 43, 446–452.

Bansal H, Kaur S P, Gupta A K (2011), 'Microsphere: Methods of preparation and applications, a comparative study', Int J P Pharm Sci Rev Res, 10, 69-78.

Bansal S, Kashyap C P, Aggarwal G, Harikumar S L (2012), 'A comparative review on vesicular drug delivery system and stability issues', Int J Res Pharm Chem, 2, 704-713.

Barbas C F, Kang A S, Lerner R A, Benkovic S J (1991), 'Assembly of combinatorial antibody libraries on phage surfaces: the gene III site', Proc Natl Acad Sci, 88, 7978–7982.

Bardonnet P L, Faivre V, Pugh W J, Piffaretti, J C, Falson. F (2006), 'Gastroretentive dosage forms: Overview and special case of Helicobacter pylori', J Control Release, 111, 1-18.

Battacharya S (2009), 'Phytosome: Emerging strategy in delivery of herbal drugs and nutraceuticals', PharmTimes, 41, 3.

Beerli R R, Rader C (2010), 'Mining human antibody repertoires', MAbs, 2, 365–378.

Bhaskaran S, Lakshmi P K (2009), 'Comparative evaluation of niosome formulations prepared by different techniques', Acta Pharm Sci, 51, 27–32.

Bochot A, Fattal E (2012), 'Liposomes for intravitreal drug delivery: a state of the art', J Control Release, 161,628–34.

Bodugöz H, Güven O (2002), 'The synthesis of nonporous poly-(isobutyl methacrylate) microspheres by suspension polymerization technique and investigation of their swelling properties', J Appl Polym Sci, 83, 349-56.

Breitz H B, Tyler A, Bjorn M J, Lesley T, Weiden P L (1997), 'Clinical experience with Tc-99m nofetumomab merpentan (Verluma) radioimmunoscintigraphy', Clinical nuclear medicine, 22, 615–20.

Burns P, Gerroir P, Mahabadi H, Patel R, Vanbesien D (2002), 'Emulsion/aggregation technology: A process for preparing microspheres of narrow polydispersity', Eur Cells Mater, 23 (Suppl. 2), 148-50.

Busby A, LaGrange L, Edwards J, King J (2002), 'The use of A Silymarin/Phospholipid compound as a fetoprotectant from ethanol induced behavioral deficits', J Herb Pharmacother, 2, 39-47.

Cao Y, You B, Wu L (2010), 'Facile Fabrication of hollow polymer microspheres through the phase inversion method', Langmuir, 26, 6115-8.

Carvalho F C, Bruschi M L, Evangelista R C, Gremião M P D (2010), 'Mucoadhesive drug delivery systems', Braz J Pharm Sci, 46, 1–17.

Cassidy J P, Landzert N M, Quadros E (1993), 'Controlled buccal delivery of buprenorphine', J Control Release, 25, 21-29.

Cevc G (1997), 'Drug delivery across the skin', Expert Opinion on Investigational Drugs, 12, 1887–937.

Chames P, Van Regenmortel M, Weiss E, Baty D (2009), 'Therapeutic antibodies: successes, limitations and hopes for the future', Brit J Pharmacol, 157, 220–233.

Cherian A K, Rana A C, Jain S K (2000), 'Self-assembled carbohydrate-stabilized ceramic nanoparticles for the parenteral delivery of insulin', Drug Dev Indust Pharm, 26, 459–469.

Clackson T, Hoogenboom H R, Griffiths A D, Winter G (1991), 'Making antibody fragments using phage display libraries', Nature, 352, 624–628.

Coelho J F, Ferreira P C, Alves P, Cordeiro R, Fonseca A C, Góis J R, Gil M H (2010), 'Drug delivery systems: Advanced technologies potentially applicable in personalized treatments', EPMA Journal, 1, 164–209.

Cui F, Cun D, Tao A, Yang M, Shi K, Zhao M, Guan Y (2005), 'Preparation and characterization of melittin-loaded poly (dl-lactic acid) or poly (dl-lactic-co-glycolic acid) microspheres made by the double emulsion method', J Control Rel, 107, 310-9.

Cummings E A, Reid G J, Finley G A, McGrath P J, Ritchie J A (1996), 'Prevalence and source of pain in pediatric inpatients', Pain, 68, 25-31

de la Fuente M, Raviña M, Paolicelli P, Sanchez A, Seijo B, Alonso M J (2010), 'Chitosan-based nanostructures: a delivery platform for ocular therapeutics', Adv Drug Deliv Rev, 62, 100–117.

Deasy P B (1984), Microencapsulation and related drug processes, In: Swarbrick J, Ed. Drugs and the Pharmaceutical Sciences. 2nd ed. New York: Marcel Dekker Inc, 20: pp. l-13.

Dhakar R C, Maurya S D, Sagar B P S, Bhagat S, Prajapati S K, Jain CP (2010), 'Variables influencing the drug entrapment efficiency of microspheres: A pharmaceutical review', Der Pharmacia Lettre, 2, 102-116.

Dhiman S, Arora S (2012), 'Niosomes A Novel drug delivery system', Int J Pharm Sci Rev Res, 15, 113-120.

Dodou D, Breedveld P, Wieringa P A (2005), 'Mucoadhesives in the gastrointestinal tract: revisiting the literature for novel applications', Eur J Pharm Biopharm, 60, 1–16.

Domingo C, Saurina J (2012), 'An overview of the analytical characterization of nanostructured drug delivery systems: towards green and sustainable pharmaceuticals: A review', Analytica Chimica Acta, 744, 8–22.

Dubey R R, Parikh J R, Parikh R R (2003), 'Effect of heating temperature and time on pharmaceutical characteristics of albumin microspheres containing 5-fluorouracil', AAPS PharmSciTech, 4, E4.

Edith M, Mark R K (1998), Encyclopedia of controlled release. London: John Wiley and Sons Inc, 2, 493-510.

Eisenbeis C F, Caligiuri M A, Byrd J C (2003), 'Rituximab: converging mechanisms of action in non-Hodgkin's lymphoma', Clin Cancer Res, 9, 5810–2.

Fiorella G, Kantor J, Aloe S, Carone M D, Spila A, D'alessandro R, Abbolito M R, Cosimelli M, Graziano F, Carboni F, Carlini S, Perri P, Sciarretta F, Greiner JW, Kashmiri SV, Steinberg S M, Roselli M, Schlom J (2001), Detection of Blood-borne Cells in Colorectal Cancer Patients by Nested Reverse Transcription-Polymerase Chain Reaction for Carcinoembryonic Antigen Messenger RNA', Cancer Res, 61, 2523–32.

Freaga M S, Salehb W M, Abdallah O Y (2018), 'Self-assembled phospholipid-based phytosomal nanocarriers as promising platforms for improving oral bioavailability of the anticancer celastrol', Int J Pharm, 535, 18–26.

Galizia G, Lieto E, De Vita F, Orditura M, Castellano P, Troiani T, Imperatore V, Ciardiello F (2007), 'Cetuximab, a chimeric human mouse anti-epidermal growth factor receptor monoclonal antibody, in the treatment of human colorectal cancer', Oncogene, 26, 3654–60.

Gandhi M, Paralkar S, Sonule M, Dabhade D, Pagar S (2014), 'Niosomes: Novel Drug Delivery System, Int J Pure App Biosci 2, 267-274.

Ganesan P, Jasmine A, Johnson D, Sabapathy L (2014), 'Review on microsphere', Am J Drug Discovery Development, 4, 153-179.

Godin B, Touitou E (2003), 'Ethosomes: new prospects in transdermal delivery', Crit Rev Ther Drug Carrier Syst, 20, 63.

Gonzales N R, de Pascalis R, Schlom J, Kashmiri S V (2005), 'Minimizing the immunogenicity of antibodies for clinical application', Tumour Biol, 26, 31-43.

Goyal A K, Rawat A, Mahor S, Gupta PN, Khatri K, Vyas S P (2006), 'Nanodecoy system: a novel approach to design hepatitis B vaccine for immunopotentiation', Int J Pharm, 309, 227–233.

Goyal K, Walton L J, Tunnacliffe A (2005), 'LEA proteins prevent protein aggregation due to water stress', Biochem J, 388(Pt 1), 151-7.

Goyal P, Goyal K, Gurusamy S, Kumar V, Singh A, Katare O P, Mishra D N (2005), 'Liposomal drug delivery systems-Clinical applications', Acta Pharm, 55, 1–25.

Gu J M, Robinson J R, Leung S H (1998), 'Binding of acrylic polymers to mucin/epithelial surfaces: structure-property relationships', Crit Rev Ther Drug Carrier Syst, 5, 21–67.

Guinedi A S, Mortada N D, Mansour S, Hathout R M (2005), 'Preparation and evaluation of reverse-phase evaporation and multilamellar niosomes as ophthalmic carriers of acetazolamide', Int J Pharm, 306, 71–82

Hafeli U (2002), 'Physics and chemistry basics of biotechnology, focus on biotechnology: Review: Radioactive microspheres for medica application, 7, 213-248.

Haghiralsadat F, Amoabediny G, Naderinezhad S, Helderd M N, Kharanaghie, E A, Zandieh-Doulabid B (2018), 'Overview of preparation methods of polymeric and lipid-based (noisome, solid lipid, liposome) nanoparticles: A comprehensive review', Int J Polym Mater Polym Biomater, 67, 383-400.

Han L, He S, Wang Y, Yang L, Liu S, Zhang T (2013), 'Advances in monoclonal antibody application in myocarditis', J Zhejiang University-Science B (Biomedicine & Biotechnology), 14, 676-687.

He P, Davis S S, Illum L (1999), 'Chitosan microspheres prepared by spray drying', Int J Pharm, 187, 53-65.

Herodin F, Thullier P, Garin D, Drouet M (2005), 'Nonhuman primates are relevant models for research in hematology, immunology and virology', Eur Cytokine Netw, 16, 104-116.

Heymann M A, Payne B D, Hoffman J I, Rudolph A M (1977), 'Blood flow measurements with radionuclide-labeled particles', Prog Cardiovas Dis, 20, 55-79.

Hiking H, Kiso Y, Wagner H, Fiebig M (1984), 'Antihepatotoxic actions of flavonolignans from Silybum marianum fruits', Planta Med, 50, 248-250.

Hong M, Zhu S, Jiang Y, Tang G, Pei Y (2009), 'Efficient tumor targeting of hydroxycamptothecin loaded PEGylated niosomes modified with transferrin', J Control Release, 133, 96-102.

Hood E, Gonzalez M, Plaas A, Strom J, Van Auker M (2007), 'Immunotargeting of nonionic surfactant vesicles to inflammation', Int J Pharm, 339, 222-30.

Hope M J, Bally M B, Mayer L D, Janoff A S, Cullis P R (1986), 'Generation of multilamellar and unilamellar phospholipid vesicles, Chem Phys Lipids, 40, 89–107.

Huang Y C, Chiang C H, Yeh M K (2003), 'Optimizing formulation factors in preparing chitosan microparticles by spray-drying method', J Microencapsul, 20, 247-60.

Hudis C A (2007), 'Trastuzumab-mechanism of action and use in clinical practice', New Engl J Med, 357, 39–51.

Jahn A, Vreeland W N, Gaitan M, Locascio L E (2004), 'Controlled vesicle self–assembly in microfluidic channels with hydrodynamic focusing', J Am Chem Soc, 126, 2674–2675.

Jain N K (2001), 'Advances in controlled and novel drug delivery system', New Delhi, CBS Publishers & Distributors, pp. 317-328.

Jain N, Gupta B P, Thakur N, Jain R, Banweer J, Jain D K, Jai S (2010), 'Phytosome: A Novel Drug Delivery System for Herbal Medicine', Int J Pharm Sci Drug Res, 2, 224-228.

Jain S S, Jagtap, P S, Dand N M, Jadhav K R, Kadam V J (2012), 'Aquasomes: A novel drug carrier', J Appl Pharm Sci, 2, 184-192

Jakobovits A, Amado R G, Yang X, Roskos L, Schwab G (2007), 'From XenoMouse technology to panitumumab, the first fully human antibody product from transgenic mice', Nat Biotechnol, 25, 1134–1143.

Jimenez-Castellanos M R, Zia H, Rhodes C T (1993), 'Mucoadhesive drug delivery systems', Drug Dev Ind Pharm, 19, 143-194.

Jose M M, Bombardelli E (1987), 'Pharmaceutical compositions containing flavanolignans and phospholipida active principles', U.S. Patent EPO209037.

Joshi A, Raje J (2002), 'Sonicated transdermal drug transport', J Control Release, 83, 13-22.

Jyothi N V, Prasanna P M, Sakarkar S N, Prabha K S, Ramaiah P S, Srawan G Y (2010), 'Microencapsulation techniques, factors influencing encapsulation efficiency', J Microencapsul, 27, 187–97.

Karimi N, Ghanbarzadeh B, Hamishehkar H, Keivani F, Pezeshki A, Gholian M M (2015), 'Phytosome and Liposome: The beneficial encapsulation systems in drug delivery and food application', Appl Food Biotechnol, 2, 17-27.

Karmakar U, Faysal M M (2009), 'Diclofenac as microspheres', The Internet J Third World Med, 8.

Kaur I P, Garg A, Singla A K, Aggarwal D (2004), 'Vesicular systems in ocular drug delivery: an overview', Int J Pharm, 269, 1–14.

Kawaguchi Y, Kono K, Mimura K, Sugai H, Akaike H, Fujii H (2007), 'Cetuximab induce antibody-dependent cellular cytotoxicity against EGFR-expressing esophageal squamous cell carcinoma', Int J Cancer, 120, 781–7.

Kawashima Y, Niwa T, Takeuchi H, Hino T, Itoh Y, Furuyama S (1991), 'Characterization of polymorphs of tranilast anhydrate and tranilast monohydrate when crystallized by two solvent change spherical crystallization techniques', J Pharm Sci, 80, 472-478.

Khan M S, Doharey V (2014), 'A review on Nasal microspheres', Int J Pharma Sci, 4, 496-506.

Khatoon M, Shah K U, Din F U, Shah S U, Rehman A U, Dilawar N, Khan A N (2017), 'Proniosomes derived niosomes: recent advancements in drug delivery and targeting', Drug Deliv, 24, 56–69

Khopade A J, Khopade S, Jain N K (2002), 'Development of haemoglobin aquasomes from spherical hydroxyapatite cores precipitated in the presence of poly(amidoamine) dendrimer', Int J Pharm, 241, 145–154.

Khutoryanskiy V V (2011), 'Advances in mucoadhesion and mucoadhesive polymers', Macromol Biosci, 11, 748–764.

Kim K K, Pack D W (2006), 'Biological and biomedical nanotechnology', Springer.

Kinloch A (1982), 'The science of adhesion', J Mater Sci, 17, 617–651.

Klang V, Matsko N B, Valenta C, Hofer F (2012), 'Electron microscopy of nanoemulsions: An essential tool for characterization and stability assessment', Micron, 43, 85–103.

Kohler G, Milstein C (1975), 'Continuous cultures of fused cells secreting antibody of predefined specificity', Nature, 256, 495–497.

Kossovsky N, A. Gelman A, E. Sponsler E, Rajguru S, Torres M, Mena E, Ly K, Festekjian A (1995b), 'Preservation of surface-dependent properties of viral antigens following immobilization on particulate ceramic delivery vehicles', J Biomed Mater Res, 29, 561–573.

Kossovsky N, Bunshah R F, Gelman A, Sponsler E D, Umarjee D M (1990), 'A non-denaturing solid phase pharmaceutical carrier comprised of surface modified nanocrystalline materials', J Appl Biomater, 1, 289-294.

Kossovsky N, Gelman A, Hnatyszyn H J, Rajguru S, Garrell RL, Torbati S, Freitas SS, Chow G M (1995a) 'Surface-modified diamond nanoparticles as antigen delivery vehicles', Bioconjug Chem, 6, 507–511.

Kossovsky N, Gelman A, Sponsler E D, Hantyszyn A J, Rajguru S, Torres M, Crowder J, Shah R (1994), 'Surface modified nanocrystalline ceramic for drug delivery applications', Biomater, 15, 1201-1207.

Kossovsky N, Gelman A, Sponsler E E (1994), 'Cross linking encapsulated haemoglobin solid phase supports: lipid enveloped haemoglobin adsorbed to surface modified ceramic particles exhibit physiological oxygen lability artif', Cells Blood Sub Biotech, 223, 479-485.

Kossovsky N, Millett D (1991), 'Materials biotechnology and blood substitutes', Mater Res Soc Bull, 16, 78-81.

Krishna K V M, Reddy C S, Srikanth S (2013), 'A review on microsphere for novel drug delivery system', Int J Res Pharm Chem, 3, 763–767.

Kumarn G P, Rajeshwarrao P (2011), 'Nonionic surfactant vesicular systems for effective drug delivery-an overview', Acta Pharmaceutica Sinica B, 1, 208–219.

La Grange L, Wang M, Watkins R, Ortiz D, Sanchez M E, Konst J, Lee C, Reyes E (1999), 'Protective effects of the flavonoids mixture, silymarin, on fetal rat brain and liver', J Ethnopharmacol, 65, 53-61.

Lefkovits J, Topol E J (1995), 'Platelet glycoprotein IIb/IIIa receptor inhibitors in ischemic heart disease', Curr Opin Cardiol, 10, 420–426.

Lijun D, Wei Q, Jingdai W, Yongrong Y, Wenqing W, Binbo J (2011), 'An improved phase-inversion process for the preparation of silica/poly[styrene-co-(acrylic acid)] core–shell microspheres: synthesis and application in the field of polyolefin catalysis', Polym Int, 60, 584-91.

Lin C Y, Lin S J, Yang Y C, Wang D Y, Cheng H F, Yeh M K (2015), 'Biodegradable polymeric microsphere based vaccine and their application in infectious disease', Human Vaccines and Immunotherapeutics, 11, 650-656.

Lin Y, Yan X Y (2004), 'Progression and direction of humanized antibody research', Chin J Biotechnol, 20, 1-5.

Linke R, Klein A, Seimetz D (2010), 'Catumaxomab: clinical development and future directions', MAbs, 2, 129–36.Liu J K H (2014), 'The history of monoclonal antibody development-Progress, remaining challenges and future innovations', Annals of Medicine and Surgery, 3, 113-116.

Liu Q, Wang L, Xiao A, Yu H, Tan Q, Ding J, Yu C (2008), 'Controllable preparation of monodisperse polystyrene microspheres with different sizes by dispersion polymerization', Macromol Symp, 261, 113-20.

Loisel S, Ohresser M, Pallardy M, Daydé D, Berthou C, Cartron G, Watier H (2007), 'Relevance, advantages and limitations of animal models used in the development of monoclonal antibodies for cancer treatment', Crit Rev Oncol/hematol, 62, 34-42.

Lokwani P, Goyal A, Gupta S, Songara R, Singh N, Rathore K (2011), 'Pharmaceutical applications of magnetic particles in drug delivery system', IJPRD, 147 – 156.

Love J C, Ronan J L, Grotenbreg G M, van der Veen A G, Ploegh H L (2006) 'A microengraving method for rapid selection of single cells producing antigen-specific antibodies', Nat Biotechnol, 24, 703-707.

Lowe D, Jermutus L (2004), 'Combinatorial protein biochemistry for therapeutics and proteomics', Cur Pharm Biotechnol, 5, 17–27.

Luciani A, Olivier J C, Clement O, Siauve N, Brillet P Y, Bessoud B, Gazeau F, Uchegbu I F, Kahn E, Frija G, Cuenod C A (2004), 'Glucose-receptor MR imaging of tumors: study in mice with PEGylated paramagnetic niosomes', Radiology, 1, 135-42.

Luo D, Han E, Belcheva N, Saltzman W M (2004), 'A self-assembled, modular delivery system mediated by silica nanoparticles', J Control Release, 95, 333-41.

Maestrelli F, Cirri M, Corti G, Mennini N, Mura P (2008), 'Development of enteric-coated calcium pectinate microspheres intended for colonic drug delivery', Eur J Pharm Biopharm, 69, 508-518.

Maiti K, Mukherjee K, Gantait A (2007), 'Curcumin–phospholipid complex, preparation, therapeutic evaluation and pharmacokinetic study in rats', Int J Pharm, 330, 155-163.

Maiti K, Mukherjee K, Gantait A, Ahamed H N, Saha B P, Mukherjee P K (2005), 'Enhanced therapeutic benefit of quercetin–phospholipid complex in carbon tetrachloride induced acute liver injury in rats: a comparative study', Iran J Pharmacol Ther, 4, 84–90.

Maloney D G, Liles T M, Czerwinski D K, Waldichuk C, Rosenberg J, Grillo-Lopez A, Levy R (1994), 'Phase I clinical trial using escalating single-dose infusion of chimeric anti-CD20 monoclonal antibody (IDEC-C2B8) in patients with recurrent B-cell lymphoma', Blood 84, 2457–66.

Manosroi A, Chutoprapat R, Abe M, Manosroi J (2008), 'Characteristics of niosomes prepared by supercritical carbon dioxide (SCCO2) fluid', Int J Pharm, 352, 248–255.

Manyak M J (2008), 'Indium-111 capromab pendetide in the management of recurrent prostate cancer', Expert Rev Anticancer Therapy, 8, 175–81.

Marianecci C, Marzio L D, Rinaldi F, Celia C, Paolino D, Alhaique F, Esposito S, Carafa M (2014), 'Niosomes from 80s to present: The state of the art', Adv Colloid and Interface Sci, 205, 187–206.

Martin P L, Weinbauer G F (2010), 'Developmental toxicity testing of biopharmaceuticals in nonhuman primates previous experience and future directions', Int J Toxicol, 29, 552-568.

Marwa A, Sammour O A, Hanaa E, Mohammed A (2013), 'Preparation and in-vitro evaluation of diclofenac sodium niosomal formulations', Int J Pharm Sci Res, 4, 1757-1765.

Mayer L D, Bally M B, Hope M J, Cultis P R (1985), 'Transmembrane pH gradient drug uptake process', Biochem Biophys Acta, 816, 294–302.

Meena K P, Dangi J S, Samal P K, Namdeo K P (2011), 'Recent advances in microspheres manufacturing technology', Int J Pharm Technol, 3, 854-893.

Mehnert W, Mäder K (2012), 'Solid lipid nanoparticles: Production, characterization and applications', Adv Drug Deliv Rev, 64, 83–101.

Mi F L, Shyu S S, Kuan C Y, Lee S T, Lu K T, Jang S F (1999), 'Chitosan polyelectrolyte complexation for the preparation of gel beads and controlled release of anti-cancer drug. Enzymatic hydrolysis of polymer', J Appl Polym Sci, 74, 1868-79.

Michael W, Gerhard W, Heinrich H, Klaush D (2010), 'Liposome preparation by single-pass process. US patent 20100316696 A1.

Milenic D E, Brady E D, Brechbiel M W (2004), 'Antibody-targeted radiation cancer therapy', Nat Rev Drug Discov, 3, 488–499.

Moebus K, Siepmann J, Bodmeier R (2009), 'Alginate–poloxamer microparticles for controlled drug delivery to mucosal tissue', Eur J Pharm Biopharm, 72, 42-53.

Moghassemi S, Hadjizadeh A (2014), 'Nano-niosomes as nanoscale drug delivery systems: An illustrated review', J Control Release, 185, 22–36

Morishita M, Peppas N A (2006), 'Is the oral route possible for peptide and protein drug delivery?', Drug Discov Today, 11, 905–10.

Mortazavi S M, Mohammadabadi M R, Khosravi-Darani K, Mozafari M R (2007), 'Preparation of liposomal gene therapy vectors by a scalable method without using volatile solvents or detergents', J Biotechnol, 129, 604–613.

Mozafari M R, Reed C J, Rostron C (2007), 'Cytotoxicity evaluation of anionic nanoliposomes and nanolipoplexes prepared by the heating method without employing volatile solvents and detergents', Pharmazie, 62, 205–209.

Mozafari M R, Reed C J, Rostron C, Hasirci V (2005), 'Are view of scanning probe microscopy investigations of liposome–DNA complexes', J Liposome Res, 15, 93–107.

Mozafari M R, Reed C J, Rostron C, Kocum C, Piskin E (2002), 'Construction of stable anionic liposome–plasmid particles using the heating method. A preliminary investigation', Cell Mol Biol Lett, 7, 923–927.

Mudshinge S R, Deore A B, Patil S, Bhalgat C M (2011), 'Nanoparticles: Emerging carriers for drug delivery. Saudi Pharm J, 19, 129–141.

Mukherjee K, Maiti K, Venkatesh M, Mukherjee P K (2008), 'Phytosome of hesperetin, a value added formulation with phytomolecules', 60th Indian Pharmaceutical Congress; New Delhi, p. 287.

Mustapha O, Din F U, Kim D W, Park J H, Woo K B, Lim S J, Youn Y S, Cho K H, Rashid R, Yousaf A M, Kim J O, Yong C S, Choi H G (2016), 'Novel piroxicam-loaded nanospheres generated by the electrospraying technique: physicochemical characterisation and oral bioavailability evaluation', J Microencapsul, 33, 323–30.

Najmuddin M, Ahmed A, Shelar S, Patel V, Khan T (2010), 'Floating microspheres of ketoprofen: Formulation and evaluation', Int J Pharm Pharmceut Sci, 2, 164-168.

Nanjwade B K, Hiremath G M, Manvi F V, Teerapon S (2013), 'Formulation and evaluation of etoposide loaded aquasomes', J Nanopharm Drug Deliv, 1, 92–101.

Naparstek E, Delukina M, Or R, Nagler A, Kapelushnik J, Varadi G, Strauss N, Cividalli G, Aker M, Brautbar C, Waldmann H, Hale G, Slavin S (1999), 'Engraftment of marrow allografts treated with Campath-1 monoclonal antibodies', Exp Hematol, 27, 1210–1218.

Nasa P, Mahant S, Sharma D (2010), 'Floating systems: A novel approach towards gastroretentive drug delivery systems', Int J Pharm Pharmceut Sci, 2, 1-7.

Neupane Y R, Srivastava M, Ahmad N, Kumar N, Bhatnagar A, Kohli K (2014), 'Lipid based nanocarrier system for the potential oral delivery of decitabine: Formulation design, characterization, ex vivo, and in vivo assessment', Int J Pharm, 477, 601–612.

Nidhi, Rashid M, Kaur V, Hallan S S, Sharma S, Mishra N (2016), 'Microparticles as controlled drug delivery carrier for the treatment of ulcerative colitis: A brief review', Saudi Pharm J, 24, 458–472.

Oviedo R I, Lopez S A D, Gasga R J, Barreda C T Q (2007), 'Elaboration and structural analysis of aquasomes loaded with indomethecin', Eur J Pharm Sci, 32, 223-230.

Pagano L, Fianchi L, Caira M, Rutella S, Leone G (2007), 'The role of Gemtuzumab Ozogamicin in the treatment of acute myeloid leukemia patients', Oncogene, 26, 3679–90.

Parhi R, Suresh P (2012), 'Preparation and Characterization of Solid Lipid Nanoparticles-a Review', Cur Drug Discov Technol, 9, 1-15.

Parhi R, Suresh P (2013), 'Supercritical fluid technology: a review', Adv Pharm Sci Technol, 1, 13-36.

Parhi R, Suresh, P, Patnaik S (2015), 'Physical means of stratum corneum barrier manipulation to enhance transdermal drug delivery', Cur Drug Deliv, 12, 122-138

Parodi B, Russo E, Caviglioli G, Cafaggi S, Bignardi G (1996), 'Development and characterization of a buccoadhesive dosage form of oxycodone hydrochloride', Drug Dev Ind Pharm, 22, 445-450.

Patel J K, Patel R P, Amin A F, Patel M M (2010), 'Bioadhesive microspheres: A review', Pharm. Rev., 4.

Patel P B, Shastri D H, Shelat P K, Shukla A K (2010), 'Ophthalmic Drug Delivery System: Challenges and Approaches', Systematic Rev Pharm, 1, 113-120.

Patil S, Pancholi S S, Agrawal S, Agrawal G P (2004), 'Surface-modified mesoporous ceramics as delivery vehicle for haemoglobin', Drug Deliv, 11, 193–199.

Paul W, Sharma C P (2001), 'Porous hydroxyapatite nanoparticles for intestinal delivery of insulin', Trends Biomater Artif Organs, 14, 37–38.

Peppas N A, Wood K M, Blanchette J O (2004), 'Hydrogels for oral delivery of therapeutic proteins', Expert Opin Biol Ther, 4, 1–7.

Peterson N C (2005), 'Advances in monoclonal antibody technology: genetic engineering of mice, cells, and immunoglobulins', ILAR J, 46, 314–319.

Rader C, Ritter G, Nathan S, Elia M, Gout I, Jungbluth A A, Cohen L S, Welt S, Old L J, Barbas C F 3rd (2000), 'The rabbit antibody repertoire as a novel source for the generation of therapeutic human antibodies', J Biol Chem, 275, 13668–13676.

Ramesh D V (2009), 'Comparison of oil-in-oil, water-in-oil-in-water and melt encapsulation techniques for the preparation of controlled release B12 poly ($\square$-caprolactone) microparticles', Trends Biomater Artif Organs, 23, 21-33.

Rashid R, Kim D W, Ud Din F, Mustapha O, Yousaf A M, Park J H, Kim J O, Yong C S, Choi H G (2015b), 'Effect of hydroxypropylcellulose and Tween 80 on physicochemical properties and bioavailability of ezetimibe-loaded solid dispersion', Carbohydr Polym, 130, 26–31.

Rashid R, Kim D W, Yousaf A M, Mustapha O, Din F U, Park JH, Yong C S, Oh Y K, Youn Y S, Kim J O, Choi H G (2015a), 'Comparative study on solid self-nanoemulsifying drug delivery and solid dispersion system for enhanced solubility and bioavailability of ezetimibe', Int J Nanomedicine, 10, 6147.

Rauta P R, Nayak B, Das S (2012), 'Immune system and immune responses in fish and their role in comparative immunity study: a model for higher organisms', Immunol let, 148, 23-33.

Ravarotto L (2004), 'Efficacy of Silymarin–Phospholipid complex in reducing the toxicity of aflatoxin B1 in broiler chicks', Poult Sci, 83, 1839-43.

Rawat M, Singh D, Saraf S, Saraf S (2008), 'Development and in vitro evaluation of alginate gel-encapsulated, chitosan-coated ceramic nanocores for oral delivery of enzyme, Drug Dev Ind Pharm, 34, 181–188.

Rawstron A C (2006), 'Immunophenotyping of plasma cells', Cur ProtocCytom, 6-23.

Raybould T J, Takahashi M (1988), 'Production of stable rabbit–mouse hybridomas that secrete rabbit mAb of defined specificity', Sci, 240, 1788–90.

Ribatti D (2014), 'From the discovery of monoclonal antibodies to their therapeutic application: An historical reappraisal', Immunol Let, 161, 96–99.

Rodgers K R, Chou R C (2016). 'Therapeutic monoclonal antibodies and derivatives: Historical perspectives and future directions', Biotechnol Adv, 34, 1149–1158.

Roopenian DC, Akilesh S (2007), 'FcRn: the neonatal Fc receptor comes of age', Nat Rev Immunol, 7, 715-725.

Roser M, Fischer D, Kissel T (1998), 'Surface-modified biodegradable albumin nano- and microspheres. II: Effect of surface charges on in vitro phagocytosis and biodistribution in rats', Eur J Pharm Biopharm, 46, 255-63.

Rupenthal I D, O'Rourke M (2016), 'Ocular drug delivery-eye on innovation', Drug Deliv and Transl Res, 6, 631–633.

Saeed A F U H, Awan S A (2016), 'Advances in Monoclonal Antibodies Production and Cancer Therapy', MOJ Immunol, 3, 1-5.

Sahil K, Akanksha M, Premjeet S, Bilandi A, Kapoor B (2011), 'Microsphere: A review', Int J Res Pharm Chem, 1, 1184-1198.

Samuni U, Navati M S, Juszczak L J, Dantsker D, Yang M, Friedman J M (2000), 'Unfolding and refolding of sol–gel encapsulated carbonmonoxymyoglobin: an orchestrated spectroscopic study of intermediates and kinetics', J Phys Chem B, 104, 10802–10813.

Sankar V, Ruckmani K, Durga S, Jailani S (2010), 'Proniosomes as drug carriers', Pak J Pharm Sci, 23, 103–107.

Sankavarapu V, Aukunuru J (2009), 'Preparation, characterization and evaluation of hepatoprotective activity of NNDMAC biodegradable parenteral sustained release microsphere', J Pharm Res Health Care, 1, 240-59.

Saraf S, Kaur C D (2010), 'Phytoconstituents as photoprotective novel cosmetic formulations', Pharmacogn Rev, 4, 1-11.

Saralidze K, Koole L H, Knetsch M L W (2010), 'Polymeric microspheres for medical applications', Mater, 3, 3537-64.

Schlachetzki F, Zhu C, Pardridge W M (2002), 'Expression of the neonatal Fc receptor (FcRn) at the blood–brain barrier', J Neurochem, 81, 203-206.

Schwitters B, Masquelier J (1993), 'OPC in practice: biflavanals and their application', Alfa Omega Rome, Italy.

Senthil A, Sivakumar T, Narayanaswamy V B (2011), 'Mucoadhesive microspheres of oral ant-diabetic drug-Glipizide using different polymers', Pharm Lett, 3, 496-506.

Shadab Md, Singh G K, Ahuja A, Khar R K, Baboota, S, Sahni J K, Ali J (2012), 'Mucoadhesive Microspheres as a Controlled Drug Delivery System for Gastroretention',Syst Rev Pharm, 3, 4-14.

Shaikh R, Singh T R R, Garland M J, Woolfson A D, Donnelly R F (2011), 'Mucoadhesive drug delivery systems', J Pharm Bioallied Sci, 3, 89–100.

Shilpa, Srinivasan B P, Chauhan M (2011), 'Niosomes as vesicular carriers for delivery of proteins and biologicals', Int J Drug Deliv, 3, 14–24.

Shim H (2016), 'Therapeutic antibodies by phage display', Cur Pharm Des, 22, 6538–59.

Singh A, Sharma P K, Malviya R (2012), 'Sustained drug delivery using mucoadhesive microspheres: The basic concept, preparation methods and recent patents', Recent Patents on Nanomed, 2, 62-77

Singh B N, Kim K H (2000), 'Floating drug delivery systems: an approach to oral controlled drug delivery via gastric retention', J Control Release, 63, 235–259.

Singh D, Rawat M S, Semalty A, Semalty M (2012). 'Rutinphospholipid complex: an innovative technique in novel drug delivery system-NDDS', Cur Drug Deliv, 9, 305-314.

Singh S, Tank N K, Dwiwedi P, Charan J, Kaur R, Sidhu P, Chugh V K (2018), 'Monoclonal Antibodies: A Review', Cur Clin Pharmacol, 13, 85-99.

Sinha V R, Agrawal M K, Kumaria R (2005), 'Influence of formulation and excipient variables on the pellet properties prepared by extrusion spheronization', Cur Drug Deliv, 2, 1-8.

Sledge Jr G W (2004), 'HERe-2 stay: the continuing importance of translational research in breast cancer', J Nat Cancer Inst, 96, 725–7.

Solanki N (2018), 'Microspheres an innovative approach in drug delivery system', MOJ Bioequiv Availab, 5, 56□58.

Song S, Tian B, Chen F, Zhang W, Pan Y, Zhang Q, Yang X, Pan W (2015), 'Potentials of proniosomes for improving the oral bioavailability of poorly water-soluble drugs', Drug Dev Ind Pharm, 41, 51–62.

Spieker-Polet H, Sethupathi P, Yam P C, Knight K L (1995), 'Rabbit monoclonal antibodies: generating a fusion partner to produce rabbit–rabbit hybridomas', Proc Natl Acad Sci, 92, 9348–52.

Srinivasan A, Mukherji S K (2011), 'Tositumomab and iodine I 131 tositumomab (Bexaar)', AJNR Am J Neuroradiol, 32, 637-8.

Straka M R, Joyce J M, Myers D T (2000), 'Tc-99m nofetumomab merpentan complements an equivocal bone scan for detecting skeletal metastatic disease from lung cancer', Clinical Nuclear Medicine, 25, 54–5.

Szczupak A, Aizik D, Moraïs S, Vazana Y, Barak Y, Bayer E A, Alfonta L (2017), 'The Electrosome: a surface-displayed enzymatic cascade in a biofuel cell's anode and a high-density surface-displayed biocathodic enzyme', Nanomater, 7, 153.

Terzano C, Allegra L, Alhaique F, Marianecci C, Carafa M (2005), 'Non-phospholipid vesicles for pulmonary glucocorticoid delivery', Eur J Pharm Biopharm, 59, 57–6.

Thanou M, Nihot M T, Jansen M, Verhoef J C, Junginger H E (2001), 'Mano-N-carboxymethyl chitosan (MCC), a polyampholytic chitosan derivative, enhances the intestinal absorption of low molecular weight heparin across intestinal epithelia in vitro and in vivo', J Pharm Sci, 90, 38-46.

Tiller T (2011), 'Single B cell antibody technologies', N Biotechnol, 28, 453–457.

Trivedi P, Verma A, Garud N (2008), 'Preparation and characterization of aceclofenac microspheres', Asian J Pharm, 2, 110-115.

Uchegbu F I, Vyas S P (1998), 'Non-ionic surfactant based vesicles (niosomes) in drug delivery', Int J Pharm, 172, 33–70.

Ud Din F, Kim D W, Choi J Y, Thapa R K, Mustapha O, Kim D S, Oh Y K, Ku S K, Youn Y S, Oh K T, Yong C S, Kim J O, Choi H G (2017), 'Irinotecan-loaded double-reversible thermogel with improved antitumor efficacy without initial burst effect and toxicity for intramuscular administration', Acta Biomater, 54, 239–48.

Ud Din F, Mustapha O, Kim D W, Rashid R, Park J H, Choi J Y, Ku S K, Yong C S, Kim J O, Choi H G (2015a), 'Novel dual-reverse thermosensitive solid lipid nanoparticle-loaded hydrogel for rectal administration of flurbiprofen with improved bioavailability and reduced initial burst effect', Eur J Pharm Biopharm, 94, 64–72.

Ud Din F, Rashid R, Mustapha O, Kim D W, Park J H, Ku S K, Oh Y-K, Kim J O, Youn Y S, Yong C S, Choi H-G (2015b), 'Development of a novel solid lipid nanoparticles-loaded dual-reverse thermosensitive nanomicelle for intramuscular administration with sustained release and reduced toxicity', RSC Adv, 5, 43687–94.

Umashankar M S, Sachdeva R K, Gulati M (2010), 'Aquasomes: a promising carrier for peptides and protein delivery', Nanomedicine: Nanotechnol Biol and Medicine, 6, 419–426

Urs A V R, Kavitha K, Socken G N (2010), 'Albumin microspheres: An unique system as drug delivery carriers for non steroidal anti-inflammatory drugs (NSAIDS)', Int J Pharm Sci Rev Res, 5, 10-17.

Valenzuela A, Aspillaga M, Vial S, Guerra R (1989), 'Selectivity of silymarin on the increase of the glutathione content in different tissues of the rat', Planta Med, 55, 420-422.

Vallelado A I, López M I, Calonge M, Sánchez A, Alonso M J (2002), 'Efficacy and safety of microspheres of cyclosporin A, a new systemic formulation, to prevent corneal graft rejection in rats', Cur Eye Res 24, 39-45.

Vasir J K, Tambwekar K, Garg S (2003), 'Bioadhesive microspheres as a controlled drug delivery system', Int J Pharm, 255, 13–32.

Vays S P, Khar R K (2004), 'Targeted & controlled Drug Delivery', New Delhi, CBC Publisher & distributors, 28-30.

Vengala P, Aslam S, Subrahmanyam C V S (2013a), 'Development and in-vitro evaluation of ceramic nanoparticles of piroxicam', Lat Am J Pharm, 32, 1124–1130.

Vengala P, Dintakurthi S, Subrahmanyam C V S (2013b), 'Lactose coated ceramic nanoparticles for oral drug delivery', J Pharm Res 7, 540–545.

Verma S, Singh S K, Syan N, Mathur P, Valecha V (2010), 'Nanoparticle vesicular systems: a versatile tool for drug delivery', J Chem Pharm Res, 2, 496–509.

Virmani T, Gupta J (2017), 'Pharmaceutical application of microspheres: An approach for the treatment of various diseases', Int J Pharm Sci Res, 8, 3252-60.

Vyas S P, Goyal A K, Khatri K, Mishra N, Mehta A, Vaidya B, Tiwari S, Vyas S P (2008), 'Aquasomes-a nanoparticulate approach for the delivery of antigen', Drug Dev Ind Pharm, 34, 1297-1305.

Vyas S P, Khar R K (2002), 'Targeted and controlled drug delivery novel carrier systems. New Delhi: CBS Publishers & Distributors, pp. 417-54.

Vyas S P, Khar R K (2002), 'Targeted and controlled drug delivery novel carrier systems. New Delhi: CBS Publishers & Distributors, p. 3-37.

Wang J, Iyer S, Fielder P J, Davis J D, Deng R (2015), 'Projecting human pharmacokinetics of monoclonal antibodies from nonclinical data: comparative evaluation of prediction approaches in early drug development', Biopharm Drug Dispos, 37, 51-65.

Wang L, Hu X, Shen B, Xie Y, Shen C, Lu Y, Qi J, Yuan H, Wu W (2015), 'Enhanced stability of liposomes against solidification stress during freeze-drying and spray-drying by coating with calcium alginate', J Drug Deliv Sci Technol, 30, 163–170.

Weber J, Peng H, Rader C (2017), 'From rabbit antibody repertoires to rabbit monoclonal antibodies', Experiment Mol Medicine, 49, e305.

Wrammert J, Smith K, Miller J, Langley W A, Kokko K (2008), 'Rapid cloning of high-affinity human monoclonal antibodies against influenza virus', Nature, 453, 667-671.

Xu R (2002), 'Photon Correlation Spectroscopy-Submicron Particle Characterization, Particle Characterization: Light Scattering Methods', Dordecht, Netherlands, Kluwer Academic Publishers, pp. 223–288.

Yadav A V, Mote H H (2007), 'Development of biodegradable starch microspheres for intranasal delivery', Indian J Pharm Sci, 70, 170-174.

Yan C, Resau JH, Hewetson J, West M, Rill W L, Kende M (1994), 'Characterization and morphological analysis of protein-loaded poly(lactide-co-glycolide) microparticles prepared by water-in-oilin-water emulsion technique', J Control Rel, 1994; 32: 231-41.

Yanyu X, Yunmei S, Zhipeng C, Quineng P (2006), 'The Preparation of silybin-phospholipid complex and the study on its pharmacokinetics in rats', Int J Pharm, 307, 77-82.

Zhang Z, Liu H, Guan Q, Wang L, Yuan H (2017), 'Advances in the isolation of specific monoclonal rabbit antibodies', Frontiers in Immunol, 8, 1-5.

List of Abbreviations

DDS	:	Drug Delivery Systems
CC	:	Coulter Counter
DLS	:	Dynamic Light Scattering
PCS	:	Photon Correlation Spectroscopy
DSS	:	Doppler Shift Spectroscopy
LM	:	Light Microscopy
PLM	:	Polarized Light Microscopy
FM	:	Fluorescence Microscopy
CM	:	Confocal Microscopy
SEM	:	Scanning Electron Microscopy
TEM	:	Transmission Electron Microscopy
AFM	:	Atomic Force Microscopy
EE	:	Entrapment Efficiency
DL	:	Drug Loading
USP	:	United States Pharmacopeia
BP	:	British Pharmacopeia
SC	:	Subcutaneous
IM	:	Intramuscular
IV	:	Intravenous
Ig	:	Immunoglobulin
mAbs	:	Monoclonal Antibodies
Fab	:	Fragment Antigen Binding
Fc	:	Fragment of Crystallization
CDRs	:	Complementarity Determining Regions
PEG	:	Polypropylene Glycol
VH	:	Heavy Chain Variable
VL	:	Light Chain Variable
FACS	:	Fluorescence-Activated Cell Sorting

ASPCs	:	Antigen-Specific Memory B Cells or Plasma/Plasmablast Cells
RT-PCR	:	Reverse-cell Transcription-Polymerase Chain Reaction
CD	:	Cluster Differentiation
FDA	:	Food and Drug Administration
NHL	:	Non-Hodgkin's Lymphoma
HER	:	Human Epidermal Growth Factor
EGFR	:	Epidermal Growth Factor Receptor
CTLA	:	Cytotoxic T Lymphocyte Antigen
HL	:	Hodgin Lymphoma
AS	:	Ankylosing Spondylitis
RA	:	Rheumatoid Arthritis
UC	:	Ulcerative Colitis
PA	:	Psoriatic Arthritis
CD	:	Crohn's Disease
TNFα	:	Tumour Necrosis Factor
IL	:	Interleukin
SLE	:	Systemic Lupus Erythematous
PCSK	:	Proprotein Convertase Subtilism/Kexin Type 9
LDL	:	Low Density Lipoprotein
VEGF	:	Vascular Endothelial Growth Factor
RSV	:	Respiratory Syncytial Virus
ADCC	:	Antibody-Dependent Cellular Cytotoxicity
CDC	:	Complement Dependent Cytotoxicity
SLV	:	Small Unilamellar Vesicles
LUV	:	Large Unilamellar Vesicles
MLV	:	multi-Lamellar Vesicles
TFH	:	Thin Film Hydration
FT-cycle	:	Freeze–Thaw Cycle
DR	:	Dehydration And Rehydration
PBS	:	Phosphate Buffer Saline
SCF	:	Supercritical Fluid
RESS	:	Rapid Expansion of Supercritical Solution
GIT	:	Gastrointestinal Tract
HIV	:	Human Immunodeficiency Virus
BSA	:	Bovine Serum Albumin
EPV	:	Epstein-Barr Virus

4

Pulmonary Drug Delivery Systems: Aerosols, Propellents, Containers Types, Preparation and Evaluation, Intranasal Route Delivery Systems; Types, Preparation And Evaluation

Chinam Niranjan Patra[*1], Goutam Kumar Jena[1], Kahnu Charan Panigrahi[1], and Suryakanta Swain[2]

[1] Department of Pharmaceutics, Roland Institute of Pharmaceutical Sciences, Berhampur-760010, Odisha, India.

[2] Southern Institute of Medical Sciences, College of Pharmacy, Department of Pharmaceutics, Guntur-522001, Andhra Pradesh, India.

4.1   Introduction

Active pharmaceutical ingredients (APIs) are administered to the respiratory tract for the treatment and prevention of airway diseases. Drug administration by this route can result in rapid onset of action. The dose of API can be lowered when compared with oral or parenteral routes. This will result in reduced adverse effect and cost. Pulmonary route is useful for poorly absorbed drug (orally) and drugs undergoing first pass metabolism. The lung can be considered as an effective route for delivery of drugs having systemic activity because of its large surface area, abundance of blood capillaries and lower thickness of the air blood barrier (Aulton, 2009).

4.2   Anatomy of Lung

Lung is the organ in which oxygen and carbon dioxide are exchanged between blood and inhaled air. It prevents the entry of foreign particles and microorganism. It also promotes the removal of foreign particles. The respiratory tract comprises central region which includes trachea, bronchi, bronchioles, terminal and respiratory bronchioles. The respiratory peripheral region includes respiratory bronchioles and alveolar regions, although there is no clear distinction between them. The upper respiratory tract comprises nose, throat, pharynx and larynx; the lower tract comprises the trachea, bronchi, bronchioles and the alveolar regions. These contain approximately $2–6 \times 10^8$ alveoli, producing surface area of about 70–80 m^2 in an adult male. The conducting airways are lined with the ciliated epithelial cells. Insoluble particles deposited on the airways walls in this region are trapped by the mucus, swept upwards from the lungs by the beating cilia and swallowed. This is presented in Figure 4.1.

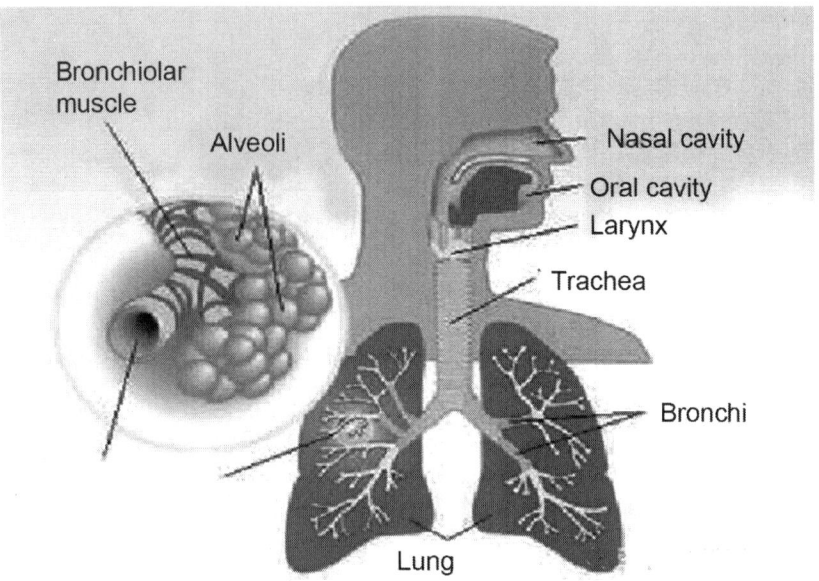

Figure 4.1: Anatomy of lung

4.3   Aerosol

Aerosolsare defined as those products that depend on the power of compressed or liquefied gas to expel the contents from the container. The products can be dispensed as fine spray or foam or semisolid stream. Aerosol can also be defined as products containing therapeutically active ingredients dissolved or emulsified in a propellant or mixture of solvent and propellant and intended for topical administration (Banker, 1990). This topical administration includes one of the following body cavities, such as ear, rectum, vagina, oral, nasal or sublingual.

4.3.1   Advantages

The product is protected from danger of contamination from outside as long as an adequate precaution is maintained within the container.

It is tamper-proof.

Metered valves aid in dispensing accurate dose.

Spray applied topically reduces irritation when compared with mechanical application of medication.

No need to the maintain sterility as required in manufacturing of parenteral.

It produces rapid onset of action.

It avoids degradation of drug product in GI tract.

Reduced dose can produce desired therapeutic effect.

The aerosol product is more convenient to use since it is a compact unit.

The aerosol dosage form is considered to be a feasible alternative when the drug exhibits erratic pharmacokinetic behaviour upon oral or parenteral administration.

4.3.2   Prerequisites for formulation of aerosol

Nonirritating to nasal mucosa.

Reasonably soluble in respiratory and nasal fluid.

Must be therapeutically effective at low dose.

Should exhibit passive transport through respiratory membrane.

Must have minimal local or topical nasal mucosal activity (unless specifically used for this purpose).

Should be stable and compatible in intranasal vehicle and pH between 5.5 and 7.5.

4.3.3   Components of aerosol package

An aerosol product consists of the following component parts: (1) propellant (2) container (3) valve and actuator and (4) product concentrate.

4.4   Propellants

The compounds which are responsible for maintaining proper pressure within the container to expel the product on opening the valve are known as propellants (Lachman et al., 2009).

4.4.1   Types of propellants

Pharmaceutical aerosols basically involve two common types of propellants, i.e. liquefied gases and compressed gases. The common examples of liquefied gases are chlorofluorocarbons (CFC), hydrocarbons, hydrochlorofluorocarbons (HCFC), hydrofluorocarbons (HFC) and the compressed gases are nitrogen, nitrous oxide and carbon dioxide (Troy, 2006).

4.4.3.1.1   Liquefied-gas propellant

When liquefied-gas propellants are sealed in an aerosol container with the product concentrate, equilibrium is established between the propellant that remains liquefied and the portion that vapourizes and occupies the head space of the container. The vapour pressure at equilibrium is characteristic for each propellant at a given temperature, and is independent of the quantity of liquefied phase present. When the aerosol canister valve is actuated, the pressure forces the liquid phase up the dip tube and out of the container. When the propellant reaches the air, it evaporates due to the drop in pressure and leaves the product concentrate as airborne liquid droplets or dry particles. As the liquid (concentrate plus propellant) is expelled from the container, the equilibrium between the liquefied and vapour phases of the propellant is rapidly reestablished. This evaporation of the liquefied propellant causes the product to cool (due to the latent heat of vapourization of the liquid propellant), resulting in a very slight pressure drop within the container. As the pressure drop is small, the product is continuously released at an even rate with the same propulsion. The product temperature (and vapour pressure) quickly returns to normal condition once the valve is released. An important characteristic of any aerosol is the density of the propellant system. Chlorofluorocarbons, HCFC, and HFC are denser than water, and will reside on the bottom of the container.Hydrocarbons are less dense than water, and will reside on top of the aqueous layer. The length of the dip tube must be considered where the concentrate and propellant reside within the can for example, with hydrocarbon propellants, the dip tube can extend through the liquid propellant to the bottom of the container.

4.4.3.1.2   Chlorofluorocarbon (CFC) propellants

Examples of CFC propellants are trichlorofluoromethane (CCl_3F, Propellant 11); dichlorodifluoromethane (CCl_2F_2; Propellant 12); and 1, 2-dichloro-1,1,2,2-tetrafluoroethane ($ClF_2C-CClF_2$, Propellant 114). These propellants are gases at room temperature that can be liquefied by cooling below their boiling point or by compressing at room temperature. These liquefied gases also have a very large expansion ratio when compared with the compressed gases (e.g. nitrogen, carbon dioxide). For many years, CFC propellants were widely used in aerosol products. However, due to their role in depleting the ozone layer of the atmosphere, their phase out and substitution with more environmentally friendly propellants started in the late 1980s under the Montreal Protocol. The US Environmental Protection Agency produces a list of acceptable propellant substitutes based on the factors, such as ozone depletion potential, global warming potential, toxicity, flammability, and exposure potential. The use of CFC propellants has now been restricted to aerosols used in the treatment of asthma and chronic obstructive pulmonary disease. Propellants 11, 12 and 114 are the choice for oral, nasal and inhalation aerosols due to their relatively low toxicity and inflammability.

4.4.3.1.3   Hydrochlorofluorocarbon (HCFC) and hydrofluorocarbon (HFC) propellants

HCFC and HFC propellants breakdown in the atmosphere at a faster rate than the CFCs, resulting in a lower ozone-depleting effect. Examples of HCFC and HFC propellants are chlorodifluoromethane ($CHClF_2$; Propellant 22); 1,1,1,2-tetrafluoroethane (CF_3CH_2F; Propellant 134a); 1-chloro-1,1-difluoroethane (CH_3CClF_2; Propellant 142b); 1,1-difluoroethane (CH_3CHF_2; Propellant 152a); and 1,1,1,2,3,3,3-heptafluoropropane (CF_3CHFCF_3; Propellant 227). Propellants 22, 142b, and 152a are used in topical pharmaceuticals. These three propellants have good miscibility with water, which makes them more useful as solvents when compared with some other propellants. Medicinal aerosols, such as asthma inhalers use the HFC propellants 134a, 227, or a combination of the two.

4.4.3.1.4   Hydrocarbon propellants

Hydrocarbon propellants are used in topical pharmaceutical aerosols because of their environmental acceptance, low toxicity, and lack of reactivity. They are useful in three-phase (two-layer) aerosol systems because they are immiscible with water and have a density less than 1. The hydrocarbons remain on top of the aqueous layer, and provide the force to expel the contents from the container. As they contain no halogens, hydrolysis does not occur making them good propellants for water-based aerosols. The main drawback is that they are flammable, and can explode. Propane (C_3H_8; Propellant A-108), butane (C_4H_{10}; Propellant A-17), and isobutane (C_4H_{10}; Propellant A-31) are the most commonly used hydrocarbons. They are used alone, as mixtures, or mixed with other liquefied gases to obtain the desired vapour pressure, density, and degree of flammability.

4.4.3.1.5   Compressed gases

In the case of compressed-gas propellants, the pressure of the compressed gas in the headspace of the aerosol container expels the product concentrate in essentially the same form as it was placed into the container. Unlike aerosols containing liquefied-gas propellants, there is no propellant reservoir. As a result, higher gas pressures are required for aerosols that use compressed gases, and the pressure within the aerosol diminishes as the product is used. Gases such as nitrogen (N_2), nitrous oxide (N_2O), and carbon dioxide (CO_2)

have been used as aerosol propellants for products that are dispensed as fine mists, foams, or semisolids, including food products, dental creams, hair preparations, and ointments. Owing to their low expansion ratio, the sprays are fairly wet, and the foams are not as stable as those produced by liquefied-gas propellants (Yung, 2014).

4.4.2   Numerical designations for fluorinated hydrocarbon propellants

The numerical designation for fluorinated hydrocarbon propellants has been designed such that the chemical structure of the compound can be determined from the number:

- The digit at the extreme right = the number of fluorine atoms;
- The second digit from the right = one more the number of hydrogen atoms, i.e. the number of hydrogen atoms plus 1;
- The third digit from the right = one less the number of carbon atoms, i.e. the number of carbon atoms minus 1. If this third digit is "0", it is omitted and a two-digit number is used;
- The capital letter "C" is used before a number to indicate the cyclic nature of a compound;
- The small letters following a number are used to indicate decreasing symmetry of isomeric compounds. The most symmetrical compound is given the designated number, and all other isomers are assigned a letter (i.e. a, b, etc.) in the descending order of symmetry;
- The number of chlorine atoms in a molecule may be determined by subtracting the total number of hydrogen and fluorine atoms from the total number of atoms required to saturate the compound.

4.5   Containers

Aerosol containers are generally made of glass, metals (e.g. tin plated steel, aluminium and stainless steel) and plastics. The selection of the container for a particular aerosol product is based on its adaptability to production methods, compatibility with the formulation, ability to sustain the pressure necessary for the product, the design and aesthetic appeal and the cost.

4.5.1   Glass containers

Glass containers would be the preferred container for most aerosols. Glass presents fewer problems with respect to chemical compatibility with the formulation when compared with metal containers and is not subject to corrosion. Glass is also more adaptive to design creativity and allows the user to view the level of contents in the container. However, glass containers must be precisely engineered to provide the maximum pressure safety and impact resistance. Therefore, glass containers are used in products that have lower pressures and lower percentages of propellants. When the pressure is below 25 psig and less than 50% propellant is used, coated glass containers are considered safe. To increase the resistance to breakage, plastic coatings are commonly applied to the outer surface of glass containers. These plastic coatings serve many purposes: (1) prevent the glass from shattering into fragments if broken; (2) absorb shock from the crimping operation during production thus decreasing the danger of breakage around the neck; (3) protect the contents from ultraviolet light; (4) act as a means of identification since the coatings are available in various colors. Glass containers range in size from 15 to 30 mL and are used primarily with solution aerosols. Glass containers are generally not used with suspension aerosols because the visibility of the suspended particles presents an aesthetic problem. All commercially available containers have a 20 mm neck finish which adapts easily to metered valves.

4.5.2   Metals

4.5.2.1   Tin-plated steel

Tin-plated steel containers are light weight and relatively inexpensive. For some products the tin provides all the necessary protection. However when required, special protective coatings are applied to the tin sheets prior to fabrication so that the inside of the container will be protected from corrosion and interaction between the tin and the formulation. The coating usually is an oleoresin, phenolic, vinyl or epoxy coating. The tin-plated steel containers are used in topical aerosols.

4.5.2.2   Aluminum

Aluminum is used in most MDIs and many topical aerosols. This material is extremely light weight and is less reactive than other metals. Aluminum containers can be coated with epoxy, vinyl or phenolic resins to decrease the interaction between the aluminum and the formulation. The aluminum can also be anodized to form a stable coating of aluminum oxide. Most aluminum containers are manufactured by an impact extrusion process that makes them seamless. Therefore, they have a greater safety against leakage, incompatibility, and corrosion.Aluminum containers are made with a 20 mm neck finish that adapts to the metered valves. For special purposes and applications, containers are also available that have neck finishes ranging from 15 to 20 mm. The container themselves are available in sizes ranging from 10 mL to over 1,000 mL (Balachandran et al., 1997).

4.5.2.3   Stainless steel

Stainless steel is used when the container must be chemically resistant to the product concentrate. The main limitation of these containers is their high cost. Plastic containers have had limited success because of their inherent permeability problems to the vapour phase inside the container. In addition, some drug–plastic interactions have limited the efficacy of the product.

4.6   Valve and Actuator

4.6.1   The valve assembly

The effectiveness of a pharmaceutical aerosol depends on achieving the proper combination of product concentrate formulation, container, and valve assembly. The valve mechanism is the part of the product package through which the contents of the container are emitted. The valve must withstand the pressure required by the product concentrate and the container, be corrosive resistant, and must contribute to the form of the emitted product concentrate. The primary purpose of the valve is to regulate the flow of product concentrate from the container. However, the valve must also be multifunctional and regulate the amount of emitted material (metered valves), be capable of delivering the product concentrate in the desired form, and be easy to turn on and off. Among the materials used in the manufacture of the various valve parts are plastic, rubber, aluminium, and stainless steel. The basic parts of a valve assembly are presented in Figure 4.2.

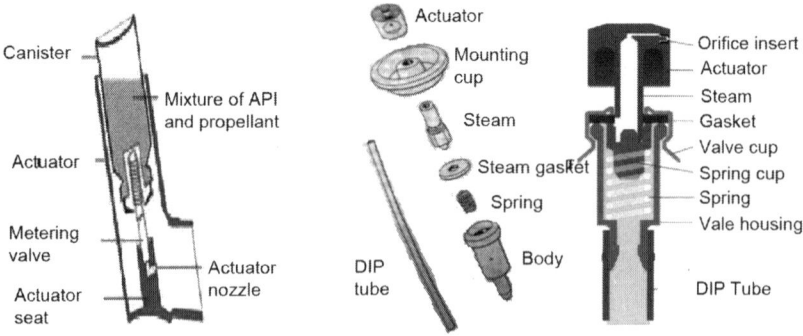

Figure 4.2: Basic components of valve assembly

4.6.1.1   Actuator

The actuator is the button which the user presses to activate the valve assembly and provides an easy mechanism of turning the valve on and off. In some actuators, mechanical breakup devices are also included. It is the combination of the type and quantity of propellant used and the actuator design and dimensions that determine the physical form of the emitted product concentrate.

4.6.1.2   Stem

The stem supports the actuator and delivers the formulation in the proper form to the chamber of the actuator.

4.6.1.3   Gasket

The gasket, placed snugly with the stem, serves to prevent leakage of the formulation of the valve is in the closed position.

Spring–the spring holds the gasket in place and also is the mechanism by which the actuator retracts when pressure is released thereby returning the valve to the closed position.

Mounting cup–the mounting cup which is attached to the aerosol container serves to hold the valve in place. Because the undersigned of the mounting cup is exposed to the formulation, it must receive the same consideration as the inner part of the container with respect to meeting criteria of compatibility. If necessary, it may be coated with an inert material to prevent an undesired interaction.

Housing–the housing located directly below the mounting cup serves as the link between the dip tube and the stem and actuator. With the stem, its orifice helps to determine the delivery rate and the form in which the product is emitted.

Dip tube–the dip tube which extends from the housing down into the product concentrate serves to bring the formulation from the container to the valve. The viscosity of the product and its intended delivery to rate dictate the inner dimensions of the dip tube and housing for a particular product.

Spray valves are used to obtain fine to coarse wet sprays. Depending on the formulation and the design of the valve and actuator, the particle size of the emitted spray can be varied. The spray is produced as an aerosol solution passes through a series of small orifices which open into chambers that allow the product concentrate to expand into the proper particle size.

Vapour tap valves are used with powder aerosols, water based aerosols, aerosols containing suspended materials, and other agents that would tend to clog a standard valve. This valve is basically a standard valve except that a small hole has been placed into the valve housing. This allows vapourized propellant to be emitted along with the product concentrate and produces a spray with greater dispersion. These valves are used with aqueous and hydroalcoholic product concentrates and hydrocarbon propellants.

Foam valves have only one orifice that leads to a single expansion chamber. The expansion chamber also serves as the delivery nozzle or applicator. The chamber is the appropriate volume to allow the product concentrate to expand into a ball of foam. Foam valves are used for viscous product concentrates, such as creams and ointments because of the large orifice and chamber. Foam valves also are used to dispense rectal and vaginal foams. If the size of the orifice and expansion chamber is appropriately reduced, a product concentrate that would produce a foam will be emitted as a solid stream. In this case, the ball of foam begins to develop where the stream impinges on a surface.

Metered dose inhaler (MDI) valves (metering valves) are used to accurately deliver a dose of medication. Metered valves are used for all oral, inhalation and nasal aerosols. The metered valves reproducibly deliver an amount of product concentrate accurately from the same package and also allow for the same accuracy between different packages (Cyr et al., 1991).The amount of material emitted is regulated by an auxiliary valve chamber of fixed capacity and dimensions. This metering chamber volume can be varied so that about 25 to 150 mL of product concentrate is delivered per actuation. Access in and out of the metering chamber is controlled by a dual valve mechanism. When the actuator is closed, a seal blocks emission from the chamber to the atmosphere. However, the chamber is open to the contents of the container and it is filled. When the actuator is depressed, the seals reverse function; the chamber becomes open to the atmosphere and releases its contents and at the same time becomes sealed from the contents of the container. When the actuator is again closed, the system prepares for the next dose. Two basic types of metering valves are available; one for inverted use and the other for upright use. Generally the valves for upright use are used with solution type aerosols and contain a thin capillary dip tube. Suspension or dispersion aerosols use the valve intended for inverted use that does not contain a dip tube. In general, valves should retain the material in the metering chamber for fairly long periods. However, it is possible for the material in the chamber to slowly return back to the container. The degree to which this occurs depends on the construction of the valve and length of time between actuations of the valves. Some valves have been fitted with a «drain tank» to overcome this problem.

4.7   Product Concentrate

4.7.1   Formulations of product concentrate of aerosol

Aerosol products consist of two components, i.e. volatile and non-volatile. The volatile portion includes single propellant or a mixture of propellant and volatile solvents whereas the non-volatile portion includes active ingredients, non-volatile solvents and dispersing agents. Types of aerosol systems are presented in Figure 4.3.

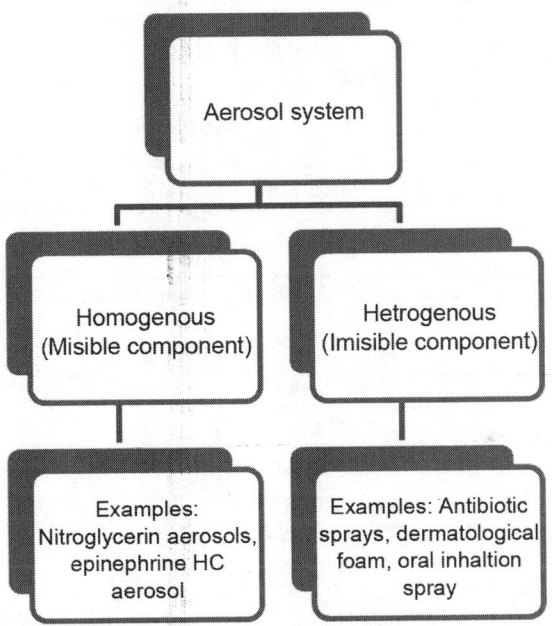

Figure 4.3: Types of aerosol system

4.8   Homogenous Aerosol Systems

4.8.1   Solution aerosol system

In this system API is dissolved in pure propellant or mixture of propellant and solvents. These are relatively easy to formulate provided the ingredients are soluble. The commonly used solvents are ethyl alcohol, polyethylene glycol, di-propyl glycol, ethyl acetate, hexylene glycol, acetone, glycol ether, etc. However, the toxicity must be given consideration. In solution aerosol system, greater the amount of propellant present, the greater will be the degree of dispersion and finer the spray. As the concentration of propellant decreases, the wetness of spray will increase. In general, proportion of propellant is higher in inhalation aerosol when compared with the topical aerosol. The particle of spray can be as low as 5 to 10 μm for inhalation aerosol, whereas 50 to 100 μm for topical aerosol. Prototype formulations of pressurized aerosol solution for oral or nasal and topical use are present in Tables 4.1 and 4.2, respectively.

Table 4.1: Prototype formulation for oral, inhalation and nasal aerosol systems

Active ingredients solvent	API dissolved in system of ethyl and isopropyl alcohol, glycol, isopropyl esters and surfactant
Antioxidant	Ascorbic acid
Preservative	Methyl and propyl paraben
Propellant	Isobutane, propane/butane, propane/isobutene, Propellant 22, 152/142, 22/142

Table 4.2: Prototype formulation for topical aerosol systems

Active ingredients solvent	API dissolved in system of ethyl and isopropylalcohol, glycols
	Isopropyl esters
	Surfactants
Antioxidant	Ascorbic acid
Preservative	Methyl and propyl paraben
Propellant (s)	Isobutane
	Propane/butane
	Propane/isobutene
	Propellant 22
	152/142, 22/142
	Dimethyl ether

4.9   Heterogeneous Aerosol System

4.9.1   Suspension system

The drug substances that are insoluble in the propellant or mixture of propellant and solvent can be suspended in the propellant vehicle. When the valve is dispersed, the suspension is emitted followed by rapid vapourization of propellant leaving behind the finely dispersed active ingredients. This suspension system has been formulated for drugs, such as antiasthematic, steroid, antibiotic. Prototype formulations of pressurized aerosol suspension for oral or nasal and topical use are presented in Tables 4.3 and 4.4, respectively. The major problems associated with these systems include caking agglomeration, particle size growth and clogging of the valve.

Table 4.3: Prototype Formulation for aerosol suspensions for inhalation use

Active ingredient (s)	Micronized
Dispersing agent or surfactant	Sorbitantrioleate, lecithin and lecithin derivatives, oleyl alcohol, ethyl alcohol
Propellant (s)	12/11, 12/114, 12/114/11, 12

Table 4.4: Prototype formulation for topical aerosol suspensions

Active ingredient (s)	Pass through a 325 mesh screen
Dispersing agent or surfactant	Isopropyl myristate
	Mineral oil
	Sorbitan esters
	Polysorbates
	Glycol ethers and derivatives
Propellant (s)	12/11; 12/114 (only if exempted)
	142, 152, 22
	Dimethyl ether

Important features

Moisture content of ingredients should be kept below 200 to 300 ppm.

Particle size should be 1 to 5 µm and for topical aerosol 40 to 50 µm.

API must have sufficient solubility in body fluids.

The density of the propellant and API should be nearly equal to reduce the rate of sedimentation.

The surfactant should be nontoxic, biodegradable and minimum irritation to respiratory pathway.

4.9.2   Emulsion aerosols

Emulsion aerosols consist of active ingredients, aqueous or a nonaqueous vehicles, surfactant and propellant. Depending on the choice of ingredients, the product can be emitted as stable or quick breaking foam. When the propellant is in the internal phase, typical foam is emitted. When the propellant is in the external phase, the product is dispensed as a spray. The different types of emulsion aerosols are described below:

Aqueous stable foams: It contains propellant in the concentration of 3 to 4%. It produces a dry spray. As the concentration of propellant goes on increasing more and more, the contents are delivered in dry form.

Nonaqueous stable foams: It can be formulated through the use of various glycols, such as polyethylene glycol.

Quick breaking foams: In this system, the propellant is in the external phase. When dispensed, the product is emitted as foam which then collapse into a liquid. This type of system is especially useful for topical medication, which can be applied to limited or large areas without the use of mechanical force.

Thermal foams: These are used to produce warm foam for shaving. They were not readily accepted by consumers. It can be used for dispensing medicated foams in which application of heat is desirable.

4.10   Manufacturing of Pharmaceutical Aerosols

Two typical processes are used for filling aerosol products cold filling (Figure 4.4A) and pressure filling (Figure 4.4B) (Dsa et al., 2014).

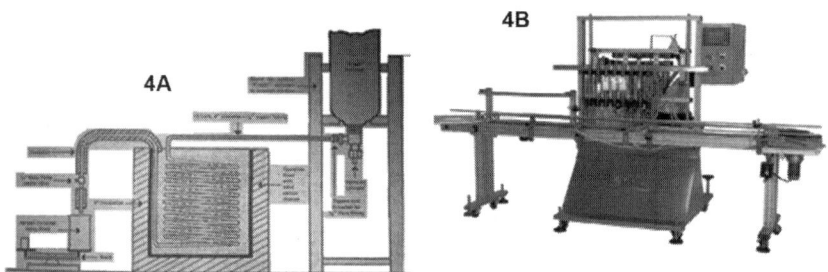

Figure 4.4: A. Cold filling apparatus, B: Pressure filling apparatus large-scale equipment

4.10.1   Cold filling

The formulation concentrate consists of a solution (or suspension) of the functional ingredient(s) dissolved in a solvent (or suspended in a carrier) that is a liquid at room temperature. In the case of drug products and MDIs, the functional ingredient would be the active pharmaceutical ingredient (API). The bulk propellant (which forms the rest of the formulation) is placed into a prechilled mixing vessel at a temperature low enough to ensure that the propellant is also in liquid form. The concentrate is added to the cold propellant in the mixing vessel, and the entire formulation is mixed to ensure homogeneity (3). The chilled-liquid formulation is then dispensed into the open aerosol canisters. The heavy vapours of the cold–liquid propellant will generally displace the air present within the canister. A valve is placed on the top of each canister and crimped into place, forming a seal between the top of the canister and a rubber gasket within the valve. Each completed aerosol is checked for weight to ensure the correct amount of formulation is in the container. Filled containers may then be passed through a heated water bath to ensure a proper seal has been formed and that there are no gaps through which the propellant may leak. The hot water bath also serves to warm the aerosol to room temperature. The formulation in the canister remains a liquid, due to the fact that it is under pressure. As the concentrate and propellant are mixed in bulk prior to being dispensed into the aerosol canister, the cold-filling process is well suited to both solution and suspension formulations including suspensions that contain a high loading of solid material. In addition, the chilling stage of the cold-filling process can control the crystallization of the product and therefore the particle size of the API.

4.10.2   Pressure filling

In contrast to cold filling, the pressure-filling process uses pressure instead of low temperature to condense the propellant. The liquid concentrate is made in the same way as for cold filling, but the propellant is held in a pressurized vessel in liquid form. There are two different methods for pressure filling the aerosol canister:

4.10.2.1   Two-stage pressure filling

The concentrate is dispensed into an open aerosol canister. A valve is then placed on top of the canister and crimped into position to form the seal. The propellant is driven under pressure through the valve and into the canister. Using this method, the mixing of the concentrate and propellant occurs in the canister, rather than in a bulk formulation tank.

4.10.2.2   Single-stage pressure filling

The concentrate and propellant are combined in a mixing vessel and held under pressure. An empty aerosol canister is assembled with a valve that is then crimped into place. The complete formulation is driven under pressure through the valve into the canister. As with the cold-filling process, the filled aerosols are checked for weight and leakage. Functionality testing of the valves at this stage also serves to rid the dip tube of pure propellant prior to consumer use. Any entrapped air in the pressure-filled package might be ignored if it does not interfere with the stability of the product, or it may be evacuated prior to or during the filling process. Pressure filling is used for most pharmaceutical aerosols. It has the following advantages: there is less danger of moisture contamination of the product (caused by condensation occurring at the low temperatures used in the cold-filling process) and also less propellant is lost in the process. Single-stage pressure filling is ideal for solutions due to the fact that the formulation can be readily driven back through the valve into the canister. Single-stage pressure filling can also be appropriate for suspensions that have a low loading of solid material. Two-stage pressure filling is generally used with suspension formulations in which the loading of the solid material makes the formulation too thick for it to be dispensed into the can through the valve with repeatable accuracy.

The filling of pharmaceutical aerosols must be conducted under sterilization as per good pharmaceutical manufacturing practice requirement. Special equipments only meant for aerosols are concentrate filler, valve placer, purger and vacuum crimper, pressure filler and leak test tank.

Manufacturing procedure

The general manufacturing procedure of pharmaceutical aerosol consists of two stages:

Stage 1: Manufacture of concentrate: The aerosol concentrate is prepared according to generally accepted procedures, and the sample is tested. Testing during this stage is done to save the time and money for verifying whether the concentrate acceptable or unacceptable.

Stage 2: Addition of propellants: After preparation of concentrate propellants are added. Once the propellant is added and the product is sealed into the container with a valve, complete rejects must be discarded. Early detection prevents loss of the other components. This would also have made it possible to correct the rejected batch instead of discarding it.

4.11   Testing of Pharmaceutical Aerosols

Testing of aerosols are divided into two categories, such as quality control of pharmaceutical aerosols and testing of pharmaceutical aerosols (3). The details are presented in Table 4.5.

Table 4.5: Testing of pharmaceutical aerosols

Quality control for pharmaceutical aerosols	Testing of pharmaceutical aerosols
a. Propellants	**A. Flammability and combustibility**
b. Valves, actuators and dip tubes	a. Flash point
c. Valve acceptance	b. Flame extension, including flashback
d. Containers	**B. Physicochemical characteristics**
e. Weight checking	a. vapour pressure
f. Leak testing	b. density
g. Spray testing	c. moisture content
	d. identification of propellant (s)
	e. Concentrate propellant ratio
	C. Performance
	a. Aerosol valve discharge rate
	b. Spray pattern
	c. Dosage with metered valve
	d. Net content
	e. Foam stability
	f. Particle size determination
	g. Leakage
	D. Biological characteristics

4.11.1   Quality control for pharmaceutical aerosols

Propellants

A sample is taken out and vapour pressure is determined. All propellants are accompanied by specification sheet. Identification of propellant can be conducted by gas chromatography. Purity of the propellant can be measured by the presence of moisture, halogen and nonvolatilevolatile residue are determined.

Valves, actuator and dip tubes

The object of this test is to determine the magnitude of valve delivery and degree of uniformity between individual valves. Standard test solutions were proposed to rule out the variation in valve delivery.Take 25 valves and placed on suitable containers. The containers are filled with specific test solutions. A button actuator with 0.02 inch orifice is attached to the valves. The filled containers are placed in a suitable atmosphere at a temperature of $25 \pm 10°C$. The filled containers are actuated to fullest extent for 2 seconds. This procedure is repeated for a total of 2 deliveries from each 25 test units.

Valve delivery per actuation in μL = Individual delivery weight in mg/ specific gravity of test solution.

Out of the 50 deliveries, if 4 or more deliveries are outside limitsand then valves are rejected. If 3 or more deliveries are outside limits, another 25 valves are tested. Lot is rejected if more than 1 delivery is outside specification. If 2 deliveries from 1 valve are beyond limits then another 25 valves are tested. Lot is rejected if more than 1 delivery is outside specification. The limit is ±15% for deliveries 54μL or less and ±10% for deliveries 55 to 200 μL.

Containers

Containers are examined for defects in linings. Quality control aspects include degree of conductivity of electric current as measure of exposed metals. Glass containers examined for flaws. The dimension of neck and other must be checked and weight should be determined.

Weight Checking

It is done by periodically adding empty tarred containers to filling lines which after filling with product concentrate are removed and reweighed. The same procedure is used for checking weight of the propellant.

Leak Testing

It is done by measuring the crimp's valve dimension and comparing. Final testing of valve enclosure is done by passing filled containers through the water bath.

Spray Testing

It is done to clear up dip tube of pure propellant and concentrate and to check any defects in the valve and the spray pattern.

4.11.2   Evaluation of pharmaceutical aerosols

4.11.2.1   Flammability and combustibility

Tag open cup apparatus is the standard test apparatus for measurement of flash point. The aerosol product is chilled to a temperature of about −25°F and transferred to the test apparatus. The temperature of the test liquid is increased slowly and the temperature at which the vapours ignite is taken as the flash point. Flame projection can be measured by spraying the aerosol to an open flame for about 4 second and the extension of the flame is measured with the help of a ruler.

4.11.2.2   Physicochemical characteristics

Specific methods are employed in order to measure the different physicochemical characteristics, such as vapour pressure, density, moisture, chemical identification, etc (Table 4.6).

Table 4.6: Physicochemical characteristics of pharmaceutical aerosols

Property	Method
1. Vapour pressure	• Can puncturing device.
2. Density	• Hydrometer • Pycnometer.
3. Moisture	• Karl–Fisher method • Gas chromatography.
4. Identification	• Gas chromatography • IR spectroscopy.

4.11.2.3   Performance

Aerosol valve discharge rate

Aerosol product of known weight is discharged for specific time. By reweighing the container, the change in the weight per time dispensed is the discharge rate in gm/sec.

Spray patterns

The method involves the impingement of sprays on a piece of paper that has been treated with dye–talc mixture. It gives a record of the spray.

Aerosol valve discharge rate

An aerosol product of known weight is taken and its contents are discharged using standard apparatus for a given period of time. The container is reweighed. Then, the change in weight per time dispensed is the discharge rate. The discharge rate can also be expressed as grams per second.

Dosage with metered valves

The doses are dispensed into the solvents or onto a material that absorbs the active ingredients. The assay of the solution gives the amount of active ingredients present.

Net contents

The tarred cans are placed onto the filling line are weighed, the difference in weight is equal to the net contents. The other method is a destructive method and consists of weighing a full container and then dispensing the contents. The contents are then weighed. The difference in weight gives the amount of contents present in the container.

4.11.2.4   *Foam stability*

The life of a foam ranges from a few seconds (for quick breaking foam) to one hour or more depending on the formulation. The methods which are used to determine the foam stability includes visual evaluation, time for a given mass to penetrate the foam, time for a given rod; that is, inserted into the foam to fall and rotational viscometer.

4.11.2.5   *Particle size determination*

Cascade impact or and light scattering decay methods are used for particle size determination.

4.12   Biological Testing

4.12.1   Therapeutic activity

For inhalation aerosols, the determination of therapeutic activity is dependent on the particle size. For topical aerosols, therapeutic activity of aerosol products are determined by applying the therapeutically active ingredients topically to the test areas and the amount of therapeutically active substances absorbed is determined.

4.12.2   Toxicity study

Topical aerosols ortopically administered aerosols are checked for chilling effect or irritation in the skin. When aerosol are topically applied, thermistor probe attached to the recording thermometer are used to determine the change in skin temperature for a given period of time. For inhalation aerosols, inhalation toxicity study is done by exposing test animals to vapours sprayed from the aerosol container.

4.13   Intranasal Route: Nasal Drug Delivery System

4.13.1   General formulation issues

Nasal dosage forms must fulfil the functions of any other types of formulation. They must: Be effective

Have an acceptable safety and stability, both chemical and microbiological.

Be acceptable to the patient to ensure compliance.

4.13.2   Advantages of nasal cavity for drug delivery

A large surface area for drug absorption.

Convenience and good patient compliance.

Rapid attainment of therapeutic drug levels in the blood.

High-drug permeability, especially for lipophilic and low molecular weight drugs.

Avoidance of harsh environmental and gastrointestinal conditions.

Bypassing of hepatic first-pass metabolism.

Potential direct drug delivery to the brain along the olfactory nerves.

Direct contact site for vaccines with lymphatic tissues.

4.13.3   Types of nasal dosage form and delivery system

The final dosage form for intranasal drug delivery is selected after due consideration of range of issues *vis-à-vis* covering patient convenience, efficiency of drug delivery and formulation issues.

4.13.3.1   Nasal dosage forms

Nasal dosage forms will usually contain the drug in a liquid or powder formulation delivered by a pressurized or pump system.

4.13.3.2   Nasal drops

Nasal drops are one of the simplest and most convenient delivery systems among all formulations. The main limitation is the lack of precision in the administered dosage and the risk of contamination during use. Nasal drops can be delivered with a pipette or by a squeezy bottle. These formulations are usually recommended for the treatment of local conditions, but challenges include microbial growth, mucociliary dysfunction and nonspecific loss from the nose or down back the throat.

4.13.3.3   Nasal sprays

Nasal spray systems consist of a chamber, a piston and an operating actuator. Nasal sprays are comparatively more accurate than drops and generate precise doses (25–200 µl) per spray. Several studies have shown that nasal sprays can produce consistent doses of reproducible plume geometry. Formulation properties, such as thixotropy, surface tension and viscosity can potentially influence droplet size and dose accuracy. Other factors, such as the applied force, orifice size and design of the pump can also affect the droplet size, which can impact the nasal deposition of sprays.

4.13.3.4   Nasal gels

A gel is a soft, solid or semi-solid-like material consisting of two or more components, one of which is a liquid, present in substantial quantity. The semisolid characteristics of gels can be defined in terms of two dynamic mechanical properties: elastic modulus G' and viscous modulus G". The rheological properties of gels depend on the polymer type, concentration and physical state of the gel. They can range from viscous solutions (e.g. hypromellose, methylcellulose, xanthan gum and chitosan) to very hard, brittle gels (e.g. gellan gum, pectin and alginate). Bioadhesive polymers have shown good potential for nasal formulations and can control the rate and extent of drug release resulting in decreased frequency of drug administration and improved patient compliance. Moreover, the prolonged contact time afforded at the site of absorption can improve drug bioavailability by slowing down mucociliary movement (Wang et al., 2019).

4.13.3.5   Nasal powders

Particulate nasal dosage forms are usually prepared by simply mixing the drug substance and the excipients by spray-drying or freeze-drying of drug. Dry-powder formulations containing bioadhesive polymers for the nasal delivery of peptides and proteins. Water-insoluble cellulose derivatives and Carbopol® 934P were mixed with insulin and the powder mixture was administered nasally. The powder took up water, swelled, and established a gel with a prolonged residence time in the nasal cavity. Over the last decade, the nasal cavity has become one the promising and potentially versatile route for delivering drugs. In particular, its unique capability of extending the drug release, by passing the hepatic first-pass metabolism and direct delivery of drugs to brain holds great promise in the field of drug delivery. A growing body of evidence relating to nasal drug delivery suggests it might the used for challenging drugs that can facilitate the pharmaceutical manufacturing and drug delivery challenges. Various pharmaceutical dosage forms and their potential to be utilized for local or systemic drug administration has been discussed in their review article. It is intuitively expected that this review will help to understand and further to develop the intranasal formulations to achieve specific therapeutic objectives. However, a number of technical and practical issues, which are also highlighted in this review article, remain a hurdle to be overcome in order for the full potential to be realized (Madney et al., 2019).

4.13.4   Evaluation of nasal formulations

All the evaluation tests for nasal formulations are common as mentioned under aerosols. Apart from the routine test, formulations should also be evaluated for pH, viscosity, osmolarity of the solution, drug release, stability studies.

4.14   Conclusions

Pharmaceutical aerosol is a noninvasive drug delivery system, which is considered as one of the best methods in comparison to other routes of drug administration. It produces better therapeutic efficacy in the treatment of diseases, such as asthma, chronic obstructive pulmonary diseases, etc. It is found to be advantageous for the following reasons, such as drug targeting, avoidance of first pass metabolism, rapid onset of action and reduced side effects. Apart from this, other dosage forms, such as nasal solutions, gels, powders are gaining better patient acceptance. Hence, pulmonary route of administration can be the future delivery system.

4.15   References

Aulton, M.E. (2009), *Aulton's Pharmaceutics: The design and manufacture of medicines.* 3rd Edition, London, Churchill Livingstone Elsevier.

Balachandran, W., Machowski, W., Gaura, E. and Hudson, C. (1997), Control of drug aerosol in human airways using electrostatic forces. *J. Electrostat.*, 40–41, 579–584.

Banker, G.S., and Rhodes, C.T. (1990), *Modern Pharmaceutics.* 2nd Edition, New York, Marcel Dekker.

Cyr, T.D., Graham, S.J. and Li, K.Y.R. (1991), Lovering EG. Low first-spray drug content in albuterol metered-dose inhalers. *Pharm. Res.* 8:658–660.

Dsa, D.J., Lechuga-Ballesteros, D. and Chan, H.K. (2014), Isothermal microcalorimetry of pressurized systems I: A rapid method to evaluate pressurized metered dose inhaler formulations. *Pharm. Res.* 31(10), 2716–2723.

Lachman, L., and Lieberman, H. A. (2009), *The Theory and Practice of Industrial Pharmacy.* Special Indian Edition, New Delhi, CBS Publishers.

Madney, Y.M., Fathy, M., Elberry, A.A., Rabea, H. and Abdelrahim, M.E. (2019), Aerosol delivery through an adult high-flow nasal cannula circuit using low-flow oxygen. *Respir. Care.* 64(4), 453-461.

Troy, D.B. (2006), *Remington: The Science and Practice of Pharmacy.* 21st Edition, Philadelphia, Lippincott Williams and Wilkins.

Wang, Q., Zuo, Z., Kit Chucky Cheung, C., Shui Yee Leung, S. (2019), Updates on thermosensitive hydrogel for nasal, ocular and cutaneous delivery. *Int. J. Pharm.* 559, 86-101.

Yung, H. (2014), Mechanisms of pharmaceutical aerosol deposition in the respiratory tract. *AAPS PharmSciTech.*15 (3), 2014.

List of Abbreviations

Active pharmaceutical ingredients = APIs

Chlorofluorocarbons = CFC

Hydrochlorofluorocarbons = HCFC
Hydrofluorocarbons = HFC
United States = US
Metered dose inhalers = MDIs
Millilitre = mL
Micrometre = μm
Parts per million = ppm
Percent = %
Degree centigrade = °C
Microliter = μL
Example = Eg

5

Nucleic Acid Based Therapeutic Delivery System: Gene Therapy, Introduction (Ex vivo and it is In vivo Gene Therapy), Potential Target Diseases for Gene Therapy (Inherited Disorder and Cancer), Gene Expression Systems (Viral and Nonviral Gene Transfer), Liposomal Gene Delivery Systems, Biodistribution and Pharmacokinetics, Knowledge of Therapeutic Antisense Molecules and Aptamers as Drugs of Future

Satya Prakash Singh[1*] and Anup Kumar Sirbaiya[1] and Suryakanta Swain[2]

[1] Pharmaceutics Division, Faculty of Pharmacy, Integral University, Certified trainer-Research Based Pedagogical Tools (IISER PUNE), Lucknow-226026, India

[2] Southern Institute of Medical Sciences, College of Pharmacy, Department of Pharmaceutics, Guntur-522001, Andhra Pradesh, India.

5.1   Introduction

Classic drug development has included a close interaction of various aspects such as biology, chemistry and pharmacology to engineer high efficacy medicines. The drawback with this approach is that the drugs are administered as such, as small molecule compounds or proteins at therapeutic concentrations, which will require re-administration at defined intervals. In contrast, if the drug can be delivered by slow release in capsules or other mechanical devices or by gene delivery vehicles as nucleic acids a more sustained form of treatment is feasible. In the latter case, these vectors can be of either viral or nonviral origin and can provide, depending on which delivery system is applied, either short- or long-term heterologous gene expression. Several key steps appear to be involved in effective gene transfer to somaticcells: (i) type of delivery vehicle that may be composed of cationic liposomes, other types of liposomes, polymers, and their combinations, various types of viral or hybrid vectors and combinations of viral vectors with polymers or lipids; (ii) interaction of the gene vehicle with serum components; (iii) its circulation time in body fluids and bio distribution; (iv) its escape from immune cells and macrophages; (v) its interaction with the surface of the cell; (vi) its triggering of apoptotic pathways by this interaction; (vii) its penetration through the cell membrane barrier; (viii) its release from endosomes or other subcellular compartments and its escape from degradation by intracellular nucleases; (ix) nuclear import; (x) ability of regulatory elements for driving the expression of the foreign gene in a particular cell type including DNA sequences that might determine integration versus episomal maintenance of a plasmid or viral vector; (xi) persistence of the plasmid in the nucleus (or of the virus) as an extra chromosomal element for many cell cycles or integration into active chromatin loci; (xii) maintenance of expression for long periods; (xiii) passage to progeny cells and (xiv) ability of the transcripts to be exported to the cytoplasm, translated, modified post-translationally and transported through the endoplasmatic reticulum and Golgi apparatus to the cell surface or extracellularly. Theoretically, a modest twofold enhancement in the efficiency of each one of the 14 steps described above would result in a 2^{14}-fold (about 16,000-fold) higher level of a therapeutic protein in the targeted cell (Culver et al., 1991, and Kay et al., 2000).

Basically gene therapy is an intracellular delivery of genomic materials (transgene) into specific cells to generate a therapeutic effect by correcting an existing abnormality or providing the cells with a new function. Different types of gene delivery systems may be applied in gene therapy to restore a specific gene function or turning off a special gene(s). The ultimate goal of gene therapy is single administration of an appropriate material to replace a defective or missing gene. The first human gene transfer was utilized in 1989 on tumour-infiltrating lymphocytes and the first gene therapy was done on ADA gene for treatment of patients with SCID (Severe Combined Immunodeficiency Defect) in 1990. Although initially the main focus of gene therapy was on inherited genetic disorders, now diverse diseases, including autosomal or X-linked recessive single gene disorders CF(Cystic Fibrosis), ADA (Adenosine Deaminase) –SCID, emphysema, retinitis pigmentosa, sickle cell anaemia, phenylketonuria, haemophilia, DMD (Duchenne Muscular Dystrophy), some autosomal dominant disorders, even polygenic disorders, different forms of cancers, vascular disease, neurodegenerative disorders, inflammatory conditions and other acquired diseases are targets of gene therapy. To date, thousands of disorders have been treated by more than hundreds of protocols of gene therapy. There are 2 major categories of gene therapy: Germline gene therapy and somatic gene therapy. Although germline gene therapy may have a great potential, because it is currently ethically forbidden, it cannot be used. To date, human gene therapy has been limited to somatic cell alterations and there is a remarkable development in the field (Cavazzana-calvo et al., 2000 and Crystal et al., 1994).

Gene therapy seeks to treat a disease by transferring one or more therapeutic nucleic acids to a patient's cells or by correcting a defective gene, for example by gene editing. Hence, this technology has the potential to cure diseases that are treatable, but not curable with conventional medications, and to provide treatments for diseases previously classified as untreatable. As with any new medical technology, translation of this concept initially led to a mixture of encouraging and disappointing results in clinical trials, and also some major setbacks. However, fueled by successful treatment of ocular diseases and primary immune deficiencies, the "comeback of gene therapy" was highlighted as one of the major scientific breakthroughs of the year by Science magazine in 2009 (Naldini 2009). Advances in the development of gene therapy vector systems, optimized for *in vivo* and *ex vivo* gene transfer, and increasing clinical experience with these technologies were major factors that have finally allowed medicine to capitalize on the potential of gene transfer for the treatment of human disease. As the field advanced gene therapy beyond correction of genetic disorders, the spectrum of applications vastly increased. In fact, eradication of blood cancers using chimeric antigen receptor (CAR)-modified T cells prompted science magazine to select cancer immunotherapy as the biggest scientific breakthrough of 2013.

Effective strategies for clinical gene therapy are based on either *in vivo* gene delivery to post-mitotic target cells or tissues or *ex vivo* gene delivery into autologous cells followed by adoptive transfer back into the patient (Herzog et al., 2010).

5.2   Types of Gene Therapy

There are two main categories of gene therapy such as somatic and germline gene therapy.

5.2.1   Germline gene therapy

Germline gene therapy alters the DNA of a sperm, ovum or fertilized ovum, so that all cells of the resultant individual will carry that change. The technology of this type of gene therapy is simple as genetic abnormalities can be corrected by direct manipulation of germline cells with no targeting, and not only achieve a cure for the individual treated, but some gametes could also carry the corrected genotype. Unlike somatic gene therapy, germline gene therapy is heritable, raising a number of medical and ethical concerns. For these reasons, germline therapy is currently not being actively pursued.

5.2.2   Somatic gene therapy

Somatic gene therapy aims to correct only the (somatic) cells of the body that are affected by the condition, i.e. in this type of gene therapy it involves the insertion of genes into diploid cells of an individual where the genetic material is not passed on to its progeny. Somatic cell therapy is viewed as a more conservative, safer approach because it affects only the targeted cells in the patient, and is not passed on to future generations; however, somatic cell therapy is short-lived because the cells of most tissues ultimately die and are replaced by new cells (Nayerossadat et al., 2012).

There are mainly two types of somatic gene therapy such as *exvivo* gene therapy and *invivo* gene therapy.

5.2.2.1 Exvivo gene therapy

In this method, some cells are taken from an appropriate organ of patient, remedial gene is introduced into the cells, and then transplanted back into that organ. The *ex vivo* gene therapy can be applied to only selected tissues whose cells can be cultured in the laboratory. Examples: bone marrow transplantation, liver transplantation, kidney transplantation, etc.

Steps involved in *ex vivo* gene therapy (Figure 5.1)
1. Isolate cells with genetic defect from a patient.
2. Grow the cells in culture.
3. Introduce the therapeutic gene to correct gene defect.
4. Select the genetically corrected cells (stable trans-formants) and grow.
5. Transplant the modified cells to the patient.

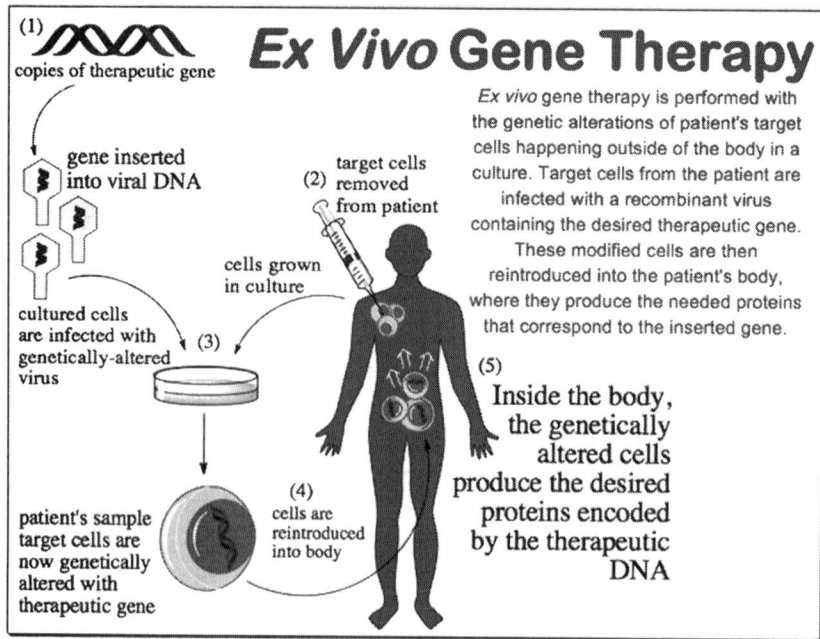

Figure 5.1: *Ex vivo* gene therapy

Example–1st gene therapy–to correct deficiency of enzyme, adenosine deaminase (ADA). It was performed on a 4-year old girl Ashanthi DeSilva, she was suffering from SCID–severe combined immunodeficiency. The SCID was caused due to the defect in gene coding for ADA.

5.2.2.2   *In vivo gene therapy*

The direct delivery of the therapeutic gene (DNA) into the target cells of a particular tissue of a patient constitutes *in vivo* gene therapy (Figure 5.2). Many tissues are the potential candidates for this approach. These include liver, muscle, skin, spleen, lung, brain and blood cells. Gene delivery can be carried out by viral or nonviral nonviral vector systems. The success of *in vivo* gene therapy mostly depends on the following parameters:

i.   The efficiency of the uptake of the remedial (therapeutic) gene by the target cells.

ii. Intracellular degradation of the gene and its uptake by nucleus.

iii. The expression capability of the gene.

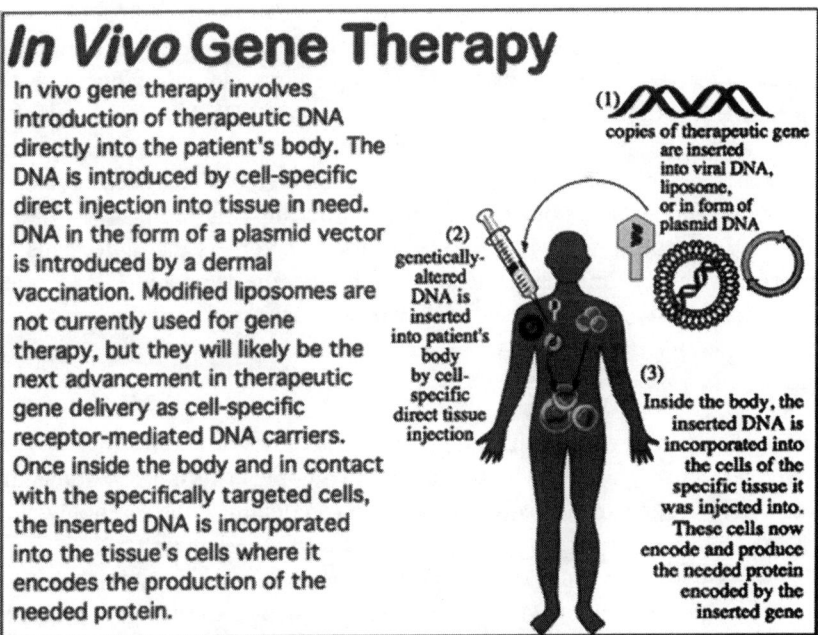

Figure 5.2: *In vivo* gene therapy

Example: In patients with cystic fibrosis, a protein called cystic fibrosis transmembrane regulator (CFTR) is absent due to a gene defect. In the absence of CFTR, chloride ions concentrate within the cells and it draws water from surroundings. This leads to the accumulation of sticky mucous in respiratory tract and lungs. Treated by *in vivo* replacement of defective gene by adenovirus vector (Crystal et al., 1994).

5.3   Different Vectors Used For Gene Delivery

1. Viral vectors
2. Nonviral vectors

5.3.1   Viral vectors

The viral vectors increase the capability of viruses to transfer genetic material into the infected cell. Viruses insert their DNA into cells with high efficiency. Vectors are evolved by genetic modification of retroviruses, adenovirus, adeno-associated virus and herpes simplex virus.

5.3.1.1   Adenoviruses

Adenoviruses (with a DNA genome) are considered to be good vectors for gene delivery because they can infect most of the nondividing human cells. The adenovirus genome comprises a double-stranded DNA molecule which is linear with a diameter of 70 mm. These are the best and mostly used systems for gene transfer. Three generations of adenoviral vectors have been developed depending on the time course of expression during viral replicative cycle. A common cold adenovirus is a frequently used vector. As the target cells are infected with a recombinant adenovirus, the therapeutic gene (DNA) enters the nucleus and expresses itself.However, this DNA does not integrate into the host genome.

5.3.1.2   Retroviruses

These are the first constructed human gene therapy vectors. Retrovirus is an enveloped virus particle, which contains two copies of viral RNA genome, rounded by a cone-shaped core. The viral RNA contains three essential genes, i.e., gag, pol and envy.

- gag → This encodes core proteins, capsid, matrix and nucleocapsid which are evolved by proteolysis cleavage of the gag precursor protein.
- pol → This encodes for viral enzyme protease, reverse transcription and integrate which are derived from gag-pol precursor.
- envy → This encodes for envelope glycoprotein which encodes envelope for glycoprotein which carry out arrival of virus.

Replication of defective retrovirus vectors that are harmless are being used. A plasmid in association with a retrovirus, a therapeutic gene and a promoter is referred to as plasmovirus. The plasmovirus is capable of carrying a DNA (therapeutic gene) of size less than 3.4 kb. As such, for the delivery of genes by retroviral vectors, the target cells must be in a dividing stage. However, the majority of the body cells are quiescent. In recent years, viral vectors have been engineered to infect nondividing cells. Further, attempts are on to include a DNA in the retroviral vectors (by engineering env gene) that encodes for cell receptor protein. If this is successfully achieved, the retroviral vector will specifically infect the target tissues (Campos and Barry 2007).

5.2.1.3   Adeno-Associated Viruses (AAV)

Adeno associated is a small virus infects humans and some other primate species, i.e., is not death causing virus which helps for immune response and also used in gene therapy and AAV viruses to the parvoviruses. Generally, it is 20 nm long and 4.5 kb size nonenveloped DNA viruses. It is a single-stranded DNA which causes latent infection to human cells. Parvoviruses gives alternative to malignancy-related retroviruses.

Structure of AAV:

1. AAV genome: Genome built of SSDNA (single-stranded deoxyribonucleic) in positive or negative sense and it consists of ITRs (Inverted terminal repeats) on both ends of SSDNA and two open reading frames (ORFs).
2. Inverted terminal repeater sequences: ITRs are named because there are symmetrical in nature and these are 145 bases each. The important property of this is it form hairpin which are useful in self-primase which are independent synthesis of the second DNA strand.
3. Rep genes: It one of the open reading frame on the left side of the genome with different lengths.
4. Cap genes: This is another reading frame we found in genome which is generally on the right hand side.
5. VP proteins: VP proteins are the part of the cap genes generally these are 3 in number named as V1, V2 and V3 (Hu and Pathak 2000, Osten et al., 2007).

5.2.1.4   Herpes simplex virus (HSV)

An enveloped, double-stranded DNA virus with the size 150 kb is herpes simplex virus. HSV is a natural human pathogen and replicating in epithelial cells. There are two members in herpes family that can infect the human body.

Gene vectors can made from the HSV in 2 methods:

1. Cloning therapeutic gene into plasmid which contains HSV and can be packed in the cells and infects with the help of HSV.
2. Cloning gene into a plasmid which was surrounded by HSV sequences and co-transfected to cells with HSV.

Advantage:

1. HSV based gene therapy has the largest cloning capacity than all the other.
2. The HSV genome is the largest of all the viral vectors.

Limitations:

HSV turns-off all the expression of gens so only small portion of HSV genome will active during latency. This can overcome by introducing foreign gene into the latency region (Goins et al., 2009).

5.3.2   Nonviral vectors

There are certain limitations in using viral vectors in gene therapy. In addition to the prohibitive cost of maintaining the viruses, the viral proteins often induce inflammatory responses in the host.

Pure DNA constructs

The direct introduction of pure DNA constructs into the target tissue is quite simple. However, the efficiency of DNA uptake by the cells and its expression are rather low. Consequently, large quantities of DNA have to be injected periodically. The therapeutic genes produce the proteins in the target cells which enter the circulation and often get degraded.

Lipoplexes

The lipid–DNA complexes are referred to as lipoplexes or more commonly liposomes. They have a DNA construct surrounded by artificial lipid layers. A large number of lipoplexes have been prepared and used. They are nontoxic and nonimmunogenic. The major limitation with the use of lipoplexes is that as the DNA is taken up by the cells, most of it gets degraded by the lysosomes. Thus, the efficiency of gene delivery by lipoplex is very low. Some clinical trials using liposome–CFTR gene complex showed that the gene expression was very short-lived (Zhang et. al., 2004).

DNA–molecular conjugates

The use of DNA–molecular conjugates avoids the lysosomal breakdown of DNA. Another advantage of using conjugates is that large-sized therapeutic

DNAs (>10 kb) can be delivered to the target tissues. The most commonly used synthetic conjugate is poly-l-lysine, bound to a specific target cell receptor. The therapeutic DNA is then made to combine with the conjugate to form a complex (Figure 5.3). This DNA molecular conjugate binds to specific cell receptor on the target cells. It is engulfed by the cell membrane to form an endosome which protects the DNA from being degraded. The DNA released from the endosome enters the nucleus where the therapeutic gene is expressed.

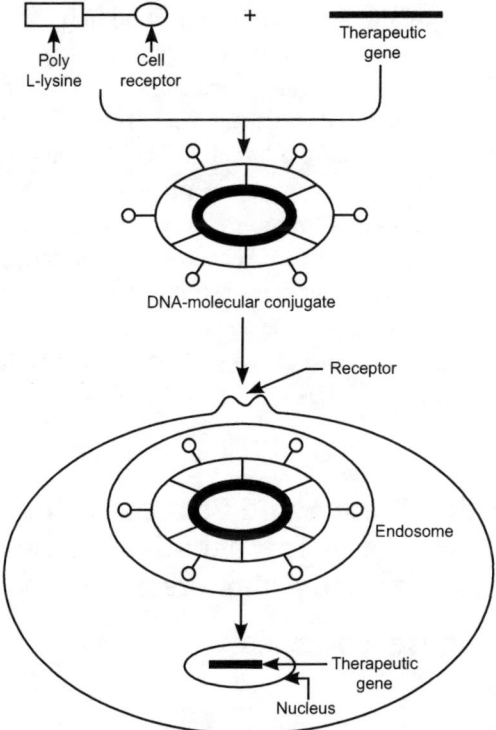

Figure 5.3: DNA–molecular conjugates

Human artificial chromosome

Human artificial chromosome (HAC) which can carry a large DNA one or more therapeutic genes with regulatory elements is a good and ideal vector. Studies conducted in cell cultures using HAC are encouraging. However, the major problem is the delivery of the large-sized chromosome into the target cells. Researchers are working to produce cells containing genetically engineered HAC. There exists a possibility of encapsulating and implanting these cells in the target tissue (Kishida et al., 2005).

5.4   Liposomal Gene Delivery Systems

Liposomal encapsulation of viruses allows systemic delivery. Liposome nanoparticle entrapment of a virus hides it from the immune system and makes its systemic delivery feasible without eliciting an immune response that would lead to the destruction of the viral particles-carriers of the therapeutic gene. Furthermore, using tumour-targeted liposomes the entire virus–liposome complex will be able to passively concentrate into solid tumours and metastases after intravenous injection into cancer patients. This system promises to solve major hurdles in gene therapy. The feasibility of the systems in SCID mice with human cancers as animal models has been tested using transfer of the β-galactosidase gene. In another study, adenovirus particles were encapsulated using bilamellar DOTAP: chol liposomes. These procedures should eliminate the problem of generating humoural immune responses against adenovirus particles when re administered and also limit the requirement of appropriate receptors on the target cells. Efficient encapsulation of adenovirus was demonstrated by electron microscopy and transduction of Ad-LacZ of otherwise adenovirus resistant cells was achieved. The encapsulated particles were also resistant to the neutralizing effect of human anti-adenovirus antibodies *ex vivo* and *in vivo* (Roder et al., 2003 and Zeisig et al., 2003).

5.5   Biodistribution and Pharmacokinetics, Knowledge of Therapeutic, Antisense Molecules and Aptamers as Drugs of Future

5.5.1   Antisense molecules

The biodistribution and pharmacokinetics of oligonucleotides have been predominantly focused on antisense single-stranded DNA/RNA oligonucleotides (ASOs), or double-stranded siRNA. ASOs were synthetic DNA/RNA-like oligonucleotides, which comprises of 16–21 nucleotides binds to RNA, and are highly water soluble (N50 mM). It is stable for years under refrigeration and administered in simple saline solutions. Two chemical classes of ASOs are commonly used today: (1) single-stranded ASO that modulate RNA function by several mechanisms, including degradation of the target RNA by the enzyme, RNase H, or modulate RNA intermediate metabolism, such as splicing and (2) double-stranded synthetic oligonucleotides that work through an RNA-induced silencing complex (RISC) to promote degradation of the target RNA.

5.5.1.1 Pharmacokinetic properties of antisense oligonucleotides

The primary route of administration for oligonucleotides for systemic applications is by parenteral injection, either intravenous (IV) infusion or subcutaneous (SC) injection. Following the systemic administration, phosphorothioate-modified single-stranded ASOs rapidly transfer from blood into tissues (minutes to hours). Pharmacokinetic properties of ASOs are similar across species and gender (5–9). Rapid transmission into cells is predominantly facilitated by endocytotic uptake. Once intracellular, ASOs exhibit long half-lives (2–4 weeks) and prolonged activity in suppressing or altering expression of their target RNA.

5.5.1.2 Absorption and distribution following IV/SC administration

Following SC administration, ASOs are rapidly absorbed from the injection site into the circulation with the peak plasma concentrations consistently reached within 3 to 4 h. Nearly complete absolute bioavailability has been observed for multiple ASOs after SC administration in monkeys. Following either IV or SC administration, plasma concentrations rapidly decline from peak concentrations in a multiexponential fashion—characterized by a dominant initial rapid distribution phase wherein drug transfers from circulation to tissues in minutes or a few hours, followed by a much slower terminal elimination phase (half-life of up to several weeks). The apparent terminal elimination rate observed in plasma is consistent with the slow elimination of ASOs from tissues, indicating equilibrium between post-distribution phase plasma concentrations and tissue concentrations. These compounds and their metabolites are readily filtered and excreted, resulting in much lower or negligible tissue uptake. The major systemic tissues of distribution include liver, kidney, bone marrow, adipocytes (cell body but not lipid fraction) and lymph nodes.

5.5.1.3 PK/PD: improved potency of ASOs drives antisense activity beyond the liver and kidney

The broad distribution of antisense drugs can be exploited to provide activity in numerous tissue targets outside of liver and kidney. It has been well understood that where the antisense oligonucleotide accumulates in highest concentrations (liver and kidney), good antisense activity is routinely observed. Nevertheless, the antisense activity has been shown in all tissues of distribution for antisense molecules.

5.5.1.4 Biodistribution following intrathecal administration

Antisense drugs do not cross the intact blood–brain barrier. However, the delivery of the antisense drugs in the cerebrospinal fluid (CSF) surrounding the spinal cord and brain results in broad distribution into spinal cord and brain tissue. These preclinical findings have now been translated to the clinic. Investigators first demonstrated safe and well-tolerated intrathecal administration of a 2'MOE modified second-generation ASO in SOD1 familial ALS patients. The second-generation 2'-MOE ASO targeting SOD1 was administered by intrathecal infusion over a 12-h period. Maximum concentrations were measured in CSF and plasma at the end of the infusion. CSF concentrations were well-predicted directly from preclinical animal PK data. The exposure of the drug in blood after direct administration into CSF was multiple orders of magnitude lower than exposures observed with direct administration by SC or IV administration reflective of the lower total dose administered and the partial transfer from the CNS to the systemic circulation (Richard et al., 2015).

5.5.1.5   Benefits of antisense drugs in therapeutics

Antisense molecules that mediate RNAi can be synthesized chemically in the laboratory and then introduced into cells to achieve targeted gene silencing. This opens up enormous possibilities for using these as potential drug candidates. Following are the key features highlighting the advantages of antisense technology, specifically siRNA:

- siRNA is a potent and highly specific therapeutic moiety.
- The traditional drugs have limited targets, whereas due to the completion of the human genome project, antisense agents like siRNA can be designed for unlimited disease targets.
- Most of the drugs act for the symptomatic relief of the disease by inhibiting the disease-causing factor. However, siRNA therapy does not allow the formation of disease-causing elements and hence acts to remove the root cause of the disease.
- Some diseases are caused by mutation in a single allele of the gene, and siRNA can be designed to act on that particular allele without affecting the normal allele.
- Once a carrier is approved for siRNA delivery, different mutant-specific siRNAs can be formulated without changing the carrier system.
- Cancer-like diseases are highly patient specific. Thus, siRNA can be explored as personalized therapy to benefit an individual patient.
- Antisense drugs have the potential to act as a chemosensitizing agent and can prevent multidrug resistance by blocking the resistance-causing component.

5.5.2   Aptamers

Aptamers are single-stranded DNA or RNA oligonucleotides that bind with high affinity and high specificity to various targets ranging from various ions, small organic compounds to large proteins and live cells. Possessing a specific and stable three-dimensional shape both *in vitro* and *in vivo*, aptamers are able to recognize their targets with high binding affinity and selectivity, and are thus termed "chemical antibodies". Aptamers are selected from large libraries of random oligonucleotides that can contain up to 1016 unique sequences. The process of aptamer selection, termed "Systematic Evolution of Ligands by EXponential Enrichment" (SELEX), was first developed in 1990 by Tuerk and Gold. SELEX is an iterative process that starts with a randomized library of oligonucleotides being incubated with the target molecule and involves iterative rounds of affinity selection and PCR amplification steps to screen for the highest affinity sequences. To date, several aptamers with high affinity and specificity have been identified for various targets with promising therapeutic and diagnostic potential using the SELEX method. Aptamers are considered to be strong chemical rivals of antibodies due to their inherent advantages over antibodies. When compared with antibodies: (i) aptamers can be produced using cell-free chemical synthesis and are therefore less expensive to manufacture on the mg scale, (ii) aptamers exhibit extremely low variability between batches and have better controlled post-production modification, (iii) are minimally immunogenic and (iv) are small in size.

5.5.2.1   Advantages and limitations of aptamers versus antibodies

Advantages of antibodies

- Pharmacokinetic and other systemic properties of antibodies are often sufficient to support product development;
- Large size prevents renal filtration and together with binding to neonatal Fc receptors can give extended circulating half-lives;
- Not susceptible to nuclease degradation;
- Antibody technologies are widely distributed because the early intellectual property either never existed or has expired.

Limitations of antibodies

- Antibodies are produced biologically in a process that is difficult to scale up without affecting product characteristics;
- Viral or bacterial contamination of manufacturing process can affect product quality, often immunogenic;

- Large size limits bioavailability or prevents access to many biological compartments;
- Limited ability to utilize negative selection pressure or to select against cell-surface targets not available in functional recombinant form;
- Susceptible to irreversible denaturation; limited shelf life.

Advantages of aptamers

- Aptamers are produced chemically in a readily scalable process;
- Chemical production process is not prone to viral or bacterial contamination;
- Nonimmunogenic;
- Smaller size allows more efficient entry into biological compartments;
- Able to select for and against specific targets and to select against cell–surface targets;
- Can usually be reversibly denatured, and phosphodiester bond is extremely chemically stable;

Limitations of aptamers

- Pharmacokinetic and other systemic properties are variable and often hard to predict;
- Small size makes them susceptible to renal filtration and they therefore have a shorter half-life;
- Unmodified aptamers are highly susceptible to serum degradation;
- Aptamer technologies are currently largely covered by a single intellectual property portfolio strategies to overcome aptamer limitations;
- Original intellectual property covering the SELEX (systematic evolution of ligands by exponential enrichment) technique will soon expire (Anthony et al., 2010).

5.5.2.2   Applications of aptamers

5.5.2.2.1   Antiviral aptamers

Targeting viral factors have become one of the most popular aptamer therapeutic uses, since they can interfere at any stage of the viral cycle, including entry, translation, replication, packaging and budding. The first viral target chosen for the aptamers isolation was the viral polymerase, largely due to its innate ability for interacting with nucleic acids (Sayer et al., 2002).

Virus entry: HIV targets helper T-cells by glycoprotein 120 (gp120). The use of gp120 as protein target for aptamers allows, not only for the inhibition of viral entry, but also for the specific delivery of other antiviral compounds.

Replication: Viral RNA-dependent RNA polymerases have been a common target for the identification of aptamers. These proteins usually share classical reverse transcriptase features with unique conformational and functional properties.

Protein synthesis and maturation: Many viruses have developed alternative translational mechanisms to that employed by the cellular cap-mRNAs; thus, rendering a candidate target for virus inhibition. For example, HCV genome is a single stranded, plus polarity RNA molecule containing in its 5' end an internal ribosome entry site (IRES), which directs protein synthesis by a cap-independent pathway. This region exhibits a high sequence and structure conservation rate that has been widely exploited as potential therapeutic target (Kim et al., 2004).

Encapsidation: The isolation of aptamers targeting viral core proteins is an interesting strategy for the investigation of the packaging process and the signals that govern it. For example, RNA aptamers against the nucleocapsid protein were isolated by different groups (Dey et al., 2005).

5.4.2.2.2   Aptamers in oncology

Cancer has been classically considered the result of genetic alterations affecting to essential signaling pathways, thus conferring proliferative and invasive properties to the carrier cell (Hanahan and Weinberg, 2000). These particular features exhibited by the malignant cell may be exploited by aptamers to specifically target and inhibit tumour progression. Aptamers have been developed against a variety of cancer targets, including extracellular ligands, cell surface proteins and intracellular factors (Hanahan and Weinberg, 2000).

5.5.2.2.3   Anticoagulant aptamers

Anticoagulants are a major class of pharmaceutical agents that can be used to prevent clotting events during certain clinical situations or for the treatment of cardiovascular diseases. The most commonly used reagent, heparin, may unleash serious secondary effects, such as haemorrhages, decrease in the platelet number and even allergies. Thus, the discovery of new anticoagulant drugs was a major goal during many years. In this context, aptamers appeared as a good alternative to classical drugs. Thrombin protein is a preferred target for the development of anticoagulant compounds. This factor is a serine protease with a key role in haemostasis by the activation of procoagulant factors, which greatly amplify the coagulation reaction (Jeter et al., 2004).

5.6   Preclinical and Clinical Trials

5.6.1   Antiangiogenic aptamers

Numerous pathological processes, such as tumourigenesis, ocular neovascular diseases or inflammatory events are intrinsically linked to the angiogenesis phenomenon. Hence, the control of the neovascularization events in these patients is a good alternative to interfere with the progress of the disease.

5.6.1.1   *Therapy against VEGF: Pegaptanib*

VEGF is a pro-angiogenic factor, which can be produced as four principal isoforms, being VEGF165 the most abundant in serum (Ferrara, 2004).

5.6.1.2   *Other antiangiogenic targets*

5.6.1.2.1   Aptamers against the complement system

The complement system is a part of the innate immune system, which is stimulated during sepsis to generate a series of activation products that finally unleash the cleavage of the C5 molecule, generating the anaphylatoxin C5a and C5b. This determines the recruitment of inflammatory cells and the liberation of proangiogenic factors, such as VEGF. This mechanism plays a key role in the progression of numerous pathophysiologic disorders. Interfering with the C5 activity has been largely proposed as a good strategy for the treatment of pathologies related to the vascularization and inflammatory processes. Different methods for blocking this cascade have been reported, including the use of neutralizing antibodies and pharmacological inhibitors.

5.6.1.2.2   Targeting of platelet-derived growth factor (PDGF)

Platelet-derived growth factor (PDGF) is an essential, ubiquitous mitogen and chemotactic factor required for the normal development of the heart, ear, central nervous system (CNS), kidney, eye and brain. In adults, it contributes to the normal maintenance of kidney, pancreas and CNS, and also regulates the blood vessels formation. Its overproduction in the absence of exogenous stimulation leads to several diseases, such as atherosclerosis, glomerulonephritis, renal failure, glioblastoma, medulloblastoma and fibrosis.

5.6.1.3   *Aptamers as antiproliferative agents*

5.6.1.3.1   AS1411

The antiproliferative DNA aptamer AS1411 was identified from a cell-based screening of guanosine-rich oligonucleotides that interfered with cellular propagation. The initial pool of oligonucleotides was shown to form homoquadruplex stabilized by G-quartets, the essential requirement for their interaction with the nucleolin protein on the cell plasma membrane.

5.6.1.3.2   *Spiegelmers for cancer therapies*

Many attempts have been accomplished for the development of spiegelmers targeting malignant cells. Two compounds, NOX-A12 and NOX-E36, are currently being tested in Phase I clinical trial for their potential as antiproliferative drugs by Noxxon Pharma AG. They bind chemokines ligand 12 and 2, respectively, which are mainly involved in tumour metastasis and inflammation. Both of them are conjugated to a 3¢-PEG molecule. NOXA12 is being assayed for the treatment of lymphoma, multiple myeloma and haematopoietic stem cell transplantation, whereas NOX-E36 is being administered to type 2 diabetes mellitus patients that suffer glomerulosclerosis for the treatment of tissue injury and inflammation (Girvan et al.2006).

5.7   Conclusions

Gene therapy can be used for the delivery of DNA into cells, this can be achieved by a number of different methods though the use of vectors, some of them are viral methods, some are nonviral methods and few are hybrid methods. When a disease attacks the humans and the viruses will attack the cells of the body, instead of giving medicines to the body through different routes, if we can use gene therapy for the treatment of some diseases then the cell have better chances of curing disease. Using the gene delivery system, these drugs will work on the cell and many more disease are cured, since they affect directly on the human immune system by attacking the cells.

5.8   References

Campos, S.K., and Barry, M.A. (2007), 'Current advances and future challenges in adenoviral vector biology and targeting', *Curr. Gene Ther.*, 7, 189–204.

Cavazzana-Calvo, M., Payen, E., Negre, O., Wang, G., Hehir, K. and Fusil, F. (2010), 'Transfusion independence and HMGA2 activation after gene therapy of human β-thalassaemia', *Nature,* 467, 318–322.

Crystal, R.G., McElvaney, N.G. and Rosenfeld, M.A. (1994), 'Administration of an adenovirus containing the human CFTR cDNA to the respiratory tract of individuals with cystic fibrosis', *Nat. Genet.*, 8, 42–51.

Culver, K.W., Anderson, W.F. and Blaese, R.M. (1991). 'lymphocyte gene therapy', *Hum. Gene Ther.*, 2, 107–109.

Dey, A.K., Khati, M., Tang, M., Wyatt, R., Lea, S.M. and James, W. (2005), 'An aptamer that neutralizes R5 strains of human immunodeficiency virus type 1 blocks gp120-CCR5 interaction', *J. Virol.*, 79, 13806–13810.

Ferrara, N. (2004), 'Vascular endothelial growth factor: basic science and clinical progress', *Endocr. Rev.*, 25, 581–611.

Girvan, A.C., Teng, Y., Casson, L.K., Thomas, S.D., Juliger, S., Ball, M.W., Klein, J.B., Pierce, W.M. Jr., Barve, S.S. and Bates, P.J. (2006), 'AGRO100 inhibits activation of nuclear factor-kappaB (NFkappaB) by forming a complex with NF-kappaB essential modulator (NEMO) and nucleolin', *Mol. Cancer Ther.*, 5, 1790–1799.

Goins, W.F., Goss, J.R., Chancellor, M.B., de Groat, W.C., Glorioso, J.C. and Yoshimura, N. (2009), 'Herpes simplex virus vector mediated gene delivery for the treatment of lower urinary tract pain', *Gene Ther.*, 16, 558-569.

Hanahan, D., and Weinberg, R.A. (2000), 'The hallmarks of cancer', *Cell*, 100, 57–70.

Herzog, R.W., Cao, O. and Srivastava, A. (2010), 'Two decades of clinical gene therapy–Success is finally mounting', *Discov. Med.,* 9, 105–111.

Hu, W.S., and Pathak, V.K. (2000), 'Design of retroviral vectors and helper cells for gene therapy', *Pharmacol. Rev.*, 52, 494–507.

Jeter, M.L., Ly, L.V., Fortenberry, Y.M., Whinna, H.C., White, R.R., Rusconi, C.P., Sullenger, B.A., Church, F.C. (2004), 'RNA aptamer to thrombin binds anion-binding exosite-2 and alters protease inhibition by heparin-binding serpins', *FEBS Lett.*, 568, 10–14.

Kay, M.A., Manna, C.S., Ragni, M.V., et al. (2000), 'Evidence for gene transfer and expression of factor IX in haemophilia B patients treated with an MV vector', *Nat. Genet.*, 24, 257–261.

Keefe, A.D., Pai, S. and Ellington, A. (2010), 'Aptamers as therapeutics', *Nat. Rev. Drug Dis.*, 9, 537–550.

Kim, M.Y., and Jeong, S. (2004), 'Inhibition of the functions of the nucleocapsid protein of human immunodeficiency virus-1 by an RNA aptamer. *Biochem'. Biophys. Res. Commun.*, 320, 1181–1186.

Kishida, T., Shin-Ya, M., Imanishi, J. and Mazda, O. (2005), 'Application of EBV-based artificial chromosome to genetic engineering of mammalian cells and tissues', *Micro Nano Mechatron. Hum. Sci.*, 7-9, 133-138.

Naldini, L. (2009), 'Medicine-A comeback for gene therapy', *Science,* 326, 805–806.

Osten, P., Grinevich, V. and Cetin, A. (2007), 'Viral vectors: A wide range of choices and high level services', *HEP*, 178, 177–202.

Richard, S.G., Norris, D., Rosie, Y. and Bennett, C.F. (2015), Pharmacokinetics, biodistribution and cell uptake of antisense oligonucleotides. *Adv. Drug Del. Rev.*, 87, 46–51.

Roder, G., Keil, O., Prisack, H.B., Bauerschmitz, G., Hanstein, B., Nestle-Kramling, C., Hemminki, A., Bender, H.G., Niederacher, D. and Dall, P. (2003), 'Novel cGMP liposomal vectors mediate efficient gene transfer', *Cancer Gene Ther.*, 10, 312–317.

Sayer, N., Ibrahim, J., Turner, K., Tahiri-Alaoui, A. and James, W. (2002), 'Structural characterization of a 2'F-RNA aptamer that binds a HIV-1 SU glycoprotein, gp120. *Biochem. Biophys. Res. Commun*, 293, 924–931.

Zeisig, R., Ress, A., Fichtner, I. and Walther, W. (2003), 'Lipoplexes with alkylphospholipid as new helper lipid for efficient *in vitro* and *in vivo* gene transfer in tumor therapy', *Cancer Gene Ther.*10, 302–311.

Zhang, S., Xu, Y., Wang, B., Qiao, W., Liu, D. and Li, Z. (2004), 'Cationic compounds used in lipoplexes and polyplexes for gene delivery', *J. Controlled Rel.*, 100, 165–180.

List of Abbreviations

SCID	:	Severe combined immunodeficiency defect
CF	:	Cystic fibrosis
ADA	:	Adenosine deaminase
DMD	:	Duchenne muscular dystrophy
CAR	:	Chimeric antigen receptor
CFTR	:	Cystic fibrosis transmembrane regulator
AAV	:	Adeno-associated viruses
SSDNA	:	Single-stranded deoxyribonucleic
ITRs	:	Inverted terminal repeats
ORFs	:	Open reading frames
HSV	:	Herpes simplex virus
HAC	:	Human artificial chromosome
DOTAP	:	N-[1-(2,3-Dioleoyloxy)-propyl]-N,N,N-trimethylammonium methyl-sulfate
ASOs	:	Antisense oligonucleotides
RISC	:	RNA-induced silencing complex
IV	:	Intravenous
SC	:	Subcutaneous
PK	:	Pharmacokinetic
PD	:	Pharmacodynamics
CSF	:	Cerebrospinal fluid
CNS	:	Central nervous system
2'MOE	:	2'-O-methoxy-ethyl
SOD1	:	Superoxide dismutase 1
ALS	:	Amyotrophic lateral sclerosis
siRNA	:	Small interfering ribonucleic acid
SELEX	:	Systematic evolution of ligands by exponential enrichment
gp120	:	Glycoprotein 120
IRES	:	Internal ribosome entry site
VEGF	:	Vascular endothelial growth factor
PDGF	:	Platelet derived growth factor
AS1411	:	Antinucleolin aptamer

Index

A

AAV, 141, 153
Active targeting, 2, 7, 8, 14
Actuator, 113, 118, 119, 120, 127, 131
ADA, 136, 153, 139
Antisense oligonucleotides, 145, 152, 153
Aptamers, 144, 147, 148, 149, 150, 152
Aqueous stable foams, 124
Atomic force microscopy, 54

B

Bioadhesive microspheres, 45, 104, 108
Biodistribution, 144, 145, 152
Blood–brain barrier, 14, 15, 20
Bulk polymerization, 49

C

Chimeric antibodies, 62
Chlorofluorocarbon (CFC), 114
Chromosome, 143, 152, 153
Coacervation technique, 48, 52
Cold filling, 124, 125, 126
Combustibility, 127, 128
Complementarity determining regions, 109
Compressed gases, 113, 114, 115
Controlled drug delivery systems, 41

D

Dendrimers, 5
Dermal and transdermal delivery, 79, 82
Detergent absorption method, 34
Detergent removal method (removal of nonencapsulated material), 34
Dialysis bag, 56
Dialysis method, 34
Dialysis, 25, 34
Differential scanning calorimetry (DSC), 34
Dilution method, 34
Double emulsion method, 50, 53, 97
Drug loading, 44, 54, 109
Dual targeting, 7, 9
Emulsion aerosols, 123

E

Emulsion polymerization, 49
Endocytosis, 9, 10, 11, 12, 20
Entrapment efficiency, 54, 74, 97, 109
Ethanol injection method, 33
Ether injection (solvent vaporization), 33
Extravasation, 12, 13, 14

F

First-order targeting, 4
Flammability, 114, 115, 127, 128

Floating microspheres, 46, 103
Fluorinated hydrocarbon propellants, 116
Foam valves, 120
Franz diffusion cell, 56
Freeze–thawed liposomes, 33
French pressure cell extrusion, 32

G

Gasket, 119, 125
Gel-permeation chromatography method, 34
Germline, 136, 137
Glass containers, 116, 117, 128

H

Heterogeneous aerosol system, 122
Hot melt microencapsulation, 48, 52
HSV, 141, 142, 153
Human antibodies, 62, 63, 64, 104
Humanized antibodies, 62, 63
Hybridoma technology, 63
Hydrocarbon propellants, 114, 115, 116, 120
Hydrochlorofluorocarbon, 113, 115, 134
Hydrofluorocarbon, 113, 115, 134

I

Immunoglobulin, 60, 104, 109
In vitro–in vivo correlations, 57
Ionic gelation method, 51, 53
Isoelectric point, 55

L

Laser diffraction, 27
Lipoplexes, 142, 152

Liposome preparation method, 32
Liposomes, 5, 17, 18, 19
Liquefied gas, 112, 113, 114, 115, 116

M

Magnetic microspheres, 47
Marker, 5
Metered dose inhaler (MDI) valves, 120
Microfluidization method, 77
Microparticles, 43, 52, 56, 57, 99, 103, 104, 105, 109
Microscopy, 26, 27, 35, 40
Microspheres and Nanoparticles, 6
Monoclonal Antibodies and Fragments, 6
Mucoadhesive microspheres, 46, 106
Murine monoclonal antibodies, 62

N

Nanocapsules, 22, 38
Nanoparticle, 37, 38, 38, 40
Nanoprecipitation, 21, 24, 38
Nanospheres, 22, 24, 25, 39
Nasal drops, 131
Nasal gels, 131
Nasal powders, 132
Nasal sprays, 131
Nonaqueous stable foams, 124

P

Particle size and size distribution, 26
Passive targeting, 2, 7, 8, 14
Percentage drug encapsulation, 35
Phase display technique, 64
Phase inversion technique, 51, 53

Phospholipid identification and assay, 36

Phytoconstituents, 42, 43, 89, 90, 91, 92, 93, 94, 106

Phytosomes or herbosomes, 89

Polymeric Micelles, 5, 6, 7, 19

Polymeric nanoparticles, 22, 29, 39, 40

Pressure filling, 124, 125, 126

Proniosomes, 74, 81, 100, 106, 107

Propellants, 113, 114, 115, 116, 117, 120, 126, 127

Q

Quick breaking foams, 124

R

Radioactive microspheres, 45, 47, 59, 98

Receptor-mediated endocytosis, 9, 11, 12

Reverse phase evaporation method, 33

Reverse phase evaporation, 75, 98

S

Salting-out, 21, 25

Scanning electron microscopy, 54, 109

SCID, 136, 139, 144, 153

SELEX, 147, 148, 153

Self-assembly, 83, 84, 99

Single B cell amplification, 64

Single emulsion method, 50, 53

Solid lipid nanoparticles, 22

Solution aerosol system, 121

Solvent diffusion, 24, 25, 39

Solvent dispersion method, 32, 33

Solvent evaporation, 21, 23, 24

Solvent evaporation, 50, 53, 91

Somatic, 135, 136, 137, 138

Sonication, 32, 33

Spectroscopy, 26, 35

Spray congealing, 49, 52

Spray drying, 49, 52, 91, 99, 108

Spray valves, 119

Stability study of liposomes, 36

Stem, 119

Supercritical fluid technology (SCF), 25

Supercritical fluid, 77, 104, 110

Surface charge, 31, 35

Suspension polymerization, 49, 96

T

Targeted drug delivery, 41, 44, 95

Theranostics, 80

Thermal foams, 124

Thin-film hydration technique, 75

Tin-plated steel, 117

Transmission electron microscopy, 54, 109

V

Valve, 112, 116, 114, 117, 118, 119, 120, 122, 123, 125, 126, 127, 128, 129

Vapour tap valves, 119

VEGF, 150, 153

Vesicle shape and lamellarity, 35

X

X-ray powder diffraction (XRPD), 28